# Horse of the Year

# Horse of the Year
## The Story of a Unique Horse Show

### Dorian Williams

David & Charles  Newton Abbot·London·Vancouver

Grand finale (*Lane*)

ISBN 0 7153 7258 0

© Dorian Williams 1976

All rights reserved. No part of this
publication may be reproduced,
stored in a retrieval system, or
transmitted, in any form or by any
means, electronic, mechanical,
photocopying, recording or
otherwise, without the prior
permission of David & Charles
(Publishers) Limited

Set in 10 on 12 Imprint
and printed in Great Britain
by Morrison & Gibb Limited
Edinburgh
for David & Charles (Publishers)
Limited
Brunel House Newton Abbot Devon

Published in Canada
by Douglas David & Charles Limited
1875 Welch Street North Vancouver
BC

# Contents

1. **A Night Out**  The touts—the atmosphere—precision timing—at control—heavy horses and Pony Club ponies—the final round—a unique magnetism—a tie-pin … 8

2. **How it all began**  Captain Tony Collings has an idea—it provokes a lukewarm response—a visit to Paris—Le Jumping—two enthusiastic converts—would it be viable?—Harringay?—Where's that?—a venture of faith is launched … 18

3. **Take-off**  Doubts—non-existent advance bookings—encouraging entries—overseas riders—a good team—names that were to become famous—old stagers—irksome delays—Pat Smythe—last-night enthusiasm—a substantial loss … 23

4. **Getting Into Its Stride**  Solving the teething troubles—the heavy horses—archaic rules—'against the clock'—guarantors—profits—show-class inconsistencies—judging methods—a ducal raspberry—a possible solution—Tony Collings' tragedy … 30

5. **The Personalities: a Virtue out of Necessity**  Need for variety—early displays—personalities—High and Mighty—signature tunes—Grand National winners—Arkle, Mandarin, Hyperion—a virtue out of necessity … 42

6. **Behind the Scenes**  The Show office—Wembley staff—the stable area—security, or lack of it—the exhibitors' corner—declarations—the caravan park—the outside arena—senior stewards—control—veterinary and medical officers, farriers and caterers—the collecting ring—the unsung heroes … 58

7. **'They also Serve'**  The hostesses—the VIPs—sponsors and sponsorship—the restaurant—timekeepers—accident drill—a clanger from the band—Horse of the Year Show directors of music—a probable protest … 69

8. **The Source of Inspiration**  Morning check-up—the escort—instructions for the day—the morning conference—post-mortem—details of the day—a working lunch—after the Show is over—late-night listening … 82

9 **So Long in the Saddle**  Colonel Sir Michael Picton  89
Ansell—early years—his career as a soldier—wounded
at St Valery—his flowers—call from the BSJA—the
Victory Championship—manager of the International—
knighthood—a long innings—no thought of giving up
dedication to detail—twenty-first criticism—
will to *win*—the undisputed stature—a tribute

10 **Fun and Games**  The Cat and Custard Pot  99
Sergeant Major Lee to the rescue—the Surtees
centenary—the Vale of Aylesbury Steeplechase—
punctured pride—St Cyr, von Nagel, Schultheis
Lise Hartel—Peralta produces a few problems—
Chamartin and Fischer—Oliviera, Gilhuys and more
criticism—the hackney carriages—the mounted games

11 **Jumping and TV**  The chicken or the egg—reasons  118
for its popularity—all that matters on the screen—
Helsinki success—a complicated technique—
homework—the mental strain—producer-commentator
relationship—the early days of television—political
brinksmanship—new sophistications—a driving
mishap—the annual 'must'—the first *Grandstand*

12 **Magic Moments**  Different types of competition—  132
past heroes—Alan Oliver versus Seamus Hayes—Ray
Howe's Puissance—George Jorrocks Hobbs—David
Broome and Sunsalve—Paddy MacMahon and
Penwood Forge Mill—comic cuts with Nugget—
Foxhunter and Finality's classic duel—Schockemohle
versus Broome—Broome's clean sweep

13 **'Plus ca Change—'**  'Always the same'—absent  144
friends—'Doc' Nichol, 'Handy' Hurrell, David Satow—
the course builders—Talbot Ponsonby and Charles
Stratton—from Harringay to Wembley—problems of
the move—attendance of the Queen—a personal
presentation—the royal box—tragedy and the
subsequent absence of Colonel Mike—a happy show

14 **The Fences Ahead**  What of the future?—an ad hoc  155
committee—the inner circle—'leave it to the boss'—
what of the chairman?—eventual need for a new
constitution—ingredients that have spelt success—
the man at the helm—that unique atmosphere—the
Cavalcade—'Where in this wide world?'

**Index**  166

# 1 A Night Out

*The touts—the atmosphere—precision timing—at control—heavy horses and Pony Club ponies—the final round—a unique magnetism—a tie-pin*

'Any tickets to spare, guvnor?'

Any badge-wearing official, arriving at the Horse of the Year Show, is accosted as he crosses the forecourt and approaches the south entrance. It is only half-way through the week, but the touts are out in force, as they have been since the first day. Just for interest one official asks a tout what a ticket is worth.

'Fourteen quid for Saturday' is the prompt reply uttered surreptitiously out of the side of the mouth, as the tout turns apparently innocently away: 'Cash.' Twenty pounds to a buyer—twenty-five or thirty by the end of the week. Nor, incredibly, will there be a shortage of takers. One steward on the first day of the Show has been offered £50 for his badge. Incredible—at a horse show!

Watching the crowds pour in, although it is the best part of an hour before the Show starts, and only a mid-week performance, perhaps it does not seem all that incredible. For thousands each year it is an event, an outing that cannot be missed. Coaches trundle in by the dozen, from Nottingham and Northampton, from Milton Keynes and Melton Mowbray. Cars with every seat occupied weave their way through the new buildings still going up, past cranes and bull-dozers, to the nearest car park. Sleek Rolls Royces and Jaguars, chauffer driven, deposit their elegant loads at the main entrance. Commissionaires wave this way and that; police move loitering cars or people along; everything must be kept flowing: indeed, the necessary security heightens the excitement, as do the astute cop-evading touts.

'Any tickets to spare, guvnor?'

Inside it is 'programmes! programmes!' as the crowds throng through the turnstiles past the advance booking office where a long queue of hopefuls wait to be told 'there's nothing left except for Thursday afternoon'.

Once through the doors to the stadium itself you are hit by the rumble and chatter of several thousands milling around the alleyway, meeting friends, having a drink or a snack, shopping at the many stands that flank the whole arena; admiring the glitter of trophies at Garrard's, looking at books, books and more books at the Horseman's Bookshop, enviously handling tack and saddlery beautiful enough to tempt the most penurious at Luxford and Parker, Giddens and Gibson; sampling the elegant riding wear at Weatherills and Harry Hall and, of course, Moss Bros; Beswick, Spillers, Sheepskin Products—they are all there; then enquiries of course at the British Horse Society (BHS) stand; perhaps buying tickets for a Munnings picture in aid of the Olympic fund.

However early you arrive, there is never enough time to see everything. Already from the arena comes the thump of a falling pole, the thud of a horse's hoof on the tan, the judges' bell, for although the show does not start officially until 7.00 pm there is still jumping from a qualifying section, as there has been in every spare moment of the day—jumping, or driving, or the preliminary judging of a show class—since 8.30 in the morning.

For the privileged there is the members' marquee, the sort one sees at weddings, with pastel colours billowing, lamps and windows, with its bar, and beyond the barrier some forty attractively laid tables. How anyone has time to eat a three-course meal is hard to imagine, except of course the regulars who are there all the week.

The arena party, provided by the Junior Leaders Regiment, The Royal Armoured Corps, stand to attention for the opening fanfare by the trumpeters of the Band of the Royal Corps of Transport (*Stonex*)

Mrs Stuart Smith's Whisky and Splash provide some light relief from the jumping (*Lane*)

For the majority though a single visit must suffice: nothing must be missed. Quickly, therefore, as soon as the lights on the arena have been lowered and the band has been spotlit and crashed into a welcoming march or selection, the stadium fills. Seats clatter as they are lowered. 'Programme! Programme!' 'Excuse me.' 'Fancy seeing you again!' It is exactly 6.45 pm as the band starts. Practically all the seats in the stadium are filled when, at precisely two minutes to seven, the house lights are lowered and a brilliant spotlight hits the entrance.

'Bang! Bang! Bang!—Bang! Bang! Bang!' from the big drum. The curtains at the entrance are drawn back. As the band punches out the Regimental March of the Royal Armoured Corps, the arena party from the Junior Leaders Regiment, the Royal Armoured Corps, march in, their brilliant yellow jerseys catching the light, smart as new pins in their neat blue trousers and red-topped caps. They snake their way across the tan between the fences, perfectly in step, halting almost aggressively smartly in the centre of the arena as the music bangs to a climax.

A moment's pause before, with immense panache, the trumpeters of the Band of the Royal Corps of Transport, who have followed the arena party into the ring and lined the ring, in front of the barrier, bathed in amber spotlights, sound a splendid fanfare.

As the last notes of the fanfare fade the band starts playing, softly and nostalgically, for it has been the Show's signature tune since 1949, the haunting theme of *Greensleeves*. As it comes to the end, the spotlights switch to the ring guard in his scarlet coat, brown boots and breeches and beaver hat.

'Ladies and gentlemen, we present the Horse of the Year Show for 1975.'

The arena party double to their corners. The ring guard raises his coach horn.

'We start this evening's performance with the William Hanson Trophy.'

A resounding call on the horn, the ring guard marches out, the full lights come up, the first horse enters the arena. The Show is under way.

With never a moment's delay horse follows horse, its score is announced, the bell is rung, fences are rebuilt: twenty horses in thirty-five minutes. Those observant enough, or in the know, will have noticed a tall figure standing at the back of 'control', pipe clenched between his teeth, a yellow carnation in his buttonhole. From time to time he can be seen taking his watch out of his breast pocket and running his thumb over the face of it. Those near enough can hear him checking the time with anyone near him.

'How many have gone?'

'Sixteen.'

'Keep pushing. We *must* finish this by twenty to. How many clear?'

'Four.'

'We could do with another two. Is Alan Ball here?'

The course builder, summoned to the side of the Show director, assures him that everything is under control.

'Well done! Keep it up!'

A few minutes later:

'Who's in the box?'

Colonel Sir Michael Ansell, Show director since 1949, though blind, is completely informed on every activity, in the closest contact with every department. Denied his sight, his mental vision ensures that the Show runs like clockwork.

'How many more?'

'Ten.'

'If there's another clear, tell the judges to bring in a standard.'

It is not necessary: the competition ends exactly on schedule. In little over 60 seconds the arena is completely cleared, waiting tractors rumbling in to collect the fences while poles and wings are stacked in the corners. As the tractors move down the sides of the arena the curtains at the entrance are drawn and in ride the prizewinners, relaxed, chatting, the horses on a long rein, looking

Ted Williams whose successes at the Horse of the Year Show spanned a quarter of a century (*Stonex*)

for the most part as relaxed as their riders. They line up in front of the royal box, the winner being called out as a smart NCO bears out the trophy, followed by two attractively attired girls with trays of rosettes. The gate of the royal box is opened and a VIP descends to make the presentation. From the corner, his hand on the elbow of his escort, his long stick with its elegantly carved head hanging on his arm, the tall figure of Sir Michael emerges as he crosses to add his congratulations to the winner.

Cup Final lap of honour at the end of the Pony Club Games (*Stonex*)

The astute will have spotted the red light suddenly wink on in the bandstand, to which surreptiously the band has returned just five minutes before the end of the competition. This is the cue from control for the band to play for the lap of honour, where possible something appropriate, *Men of Harlech* for David Broome, *Ilkley Moor bar Tat* for a Yorkshireman, *the* Yorkshireman, or any others from that sporting county; *Lili Marlene* for a German rider and so on. The band starting up is the cue for the steward to send the riders on their way. All but the winner immediately exit: the VIP returns to the box, the lights are lowered, the spots pick up the winner and away he goes on his lap of honour.

Even at the end of the first competition, indeed, even on the first day of the show, the packed audience claps in time to the tune, sure evidence that all feel involved, participating themselves, rather than watching dispassionately.

Scarcely is the winner out of the arena than the curtains are drawn again and the great Shire horses file in, magnificent in their tossing red plumes, the ribbons from their rosettes floating from the browbands, their drivers immaculately old-fashioned in dull red smocks and flat brown bowlers. The heavies enter to a cheer and each movement in their traditional drive brings forth renewed applause, the cartwheel evoking more clapping in time as the band swings in to *Light Cavalry*. But as the great horses surge down the arena at the end there are many whose eyes are moist, such is the emotional impact of this superb spectacle from a bygone era.

What a contrast is the next event, the Mounted Games of the Pony Club. Heralded by screams of delight from their supporters in the seats behind the bandstand, the four teams enter the arena, turning to wave as they pass the exhibitors' seats. One lap and then down to the games in earnest, the hysterical support from the youngsters quickly becoming infectious so that in minutes, in a manner surely most unusual at a dignified horse show, the entire stadium is engaged in noisy, enthusiastic, overwhelming partisanship, drowning the band, the loudspeakers and even the team managers' instructions.

So it goes on: event following event with the contrast of show

'When grace is laced with muscle' (*Stonex*)

classes and display, breathtaking horsemanship and solemn presentation. The audience becomes more and more involved, greeting everything with spontaneous applause until, with the final fanfare at the end of the Parade of Personalities which brings the first half of the Show to a close, there is a rush from the seats with a clatter and bang as everyone makes for the bars and refreshment kiosks, the stands and stalls, determined to be back in good time for the second half.

With television starting at 9.30 pm and the first half running a few minutes late the interval is reduced from twenty minutes to fifteen minutes. In no time the show jumping course is erected, course builders hurrying from one fence to another with rods and measuring tapes, arena party scurrying hither and thither with poles, shrubs, wings; judges walk determinedly across the arena to inspect this fence or that; in the middle of it all stands Sir Michael, hands on tall stick, pipe clenched between teeth. Suddenly the entrance gates are opened and the arena is flooded with scarlet coats: the riders are walking the course. One, two, three, four, five, six, seven: again and again they pace the distances in a combination; professionally they put a hand on a pole and roll it, to see how firm it is in the 'cups', they stand eyeing an angle, a turn—how tight is it safe to come in? The course builder is approached for clarification on some detail: what is the jump-off course?—the time allowance?

The bell rings urgently, authoritatively. Time to clear the ring. Most riders hurry out, one or two linger. The bell rings again, finally and the television lights blaze on. The competition is announced and the ring guard blows his horn. There is a hush as the first horse enters. This is the big moment, the main event of the evening—some thirty top-class internationals jumping for £1000 and a handsome trophy.

For the next 60 minutes jumper follows jumper, each on cue, entering from the inside collecting ring, which is now a hive of activity, staccato shouts breaking the tense silence. The constant instructions on the loudspeakers ensure that the next horse is always ready, and that those in the outside collecting ring know how far the competition has progressed.

Silently, almost eerily, the television cameras can be seen tirelessly following, from their vantage points round the arena, the horse in the ring. Millions at home are sharing the excitement of the moment with the thousands in the stadium—excitement which is heightened when it comes to the jump-off against the clock, and which reaches a climax when victory is clinched with the very last round of all by one of the favourites, David Broome, Harvey Smith, Eddie Macken, Alwyn Schockemohle, Graham Fletcher, Malcolm Pyrah, any of whose horses are as popular with the crowd as their riders.

So to the presentation again: the trophy, the VIP, the rosette girls, the lap of honour, the exit march, and finally the *National Anthem*.

It is late. Many have hurried off as soon as the competition is finished. Yet others linger seeming loath to bring to an end their visit to the Horse of the Year Show. On the last night, at the end of the Cavalcade, as many as 70 per cent of the audience remain in their seats until the very last of the 130 or so horses has left the arena, as much as ten minutes after the anthem. It is this, perhaps, as much as anything that explains the success of the Horse of the Year Show. Affection as well as enjoyment plays an important part. It is *our* Show; it belongs to *us*; we love it; we would hate to miss it; we are sad when it is all over.

Many depart happy and satiated with a wonderful evening's entertainment. Some, perhaps, are a little envious as, making their way to the car park, they hear the singing and hilarity from the jumpers and their hangers-on all clustered round the Foxhunter Bar; for many each year would like to be even more involved than they are: be even more part of the Show. There is a unique magnetism about the Horse of the Year Show. It cannot come round again too soon.

Not least for the touts.

At the end of the Show in 1975 Mike Ansell was given a small package. It contained a silver tie-pin: a horse shoe on a riding whip. It was from the ticket touts. With it was the message:

'Thanks for all you've done for us.'

# 2 How it all began

Captain Tony Collings has an idea—it provokes a lukewarm response—a visit to Paris—Le Jumping—two enthusiastic converts—would it be viable?—Harringay? Where's that?—a venture of faith

How did it all begin? During 1948, a certain Captain Tony Collings, Director at that time of the Porlock Vale Riding School, then and now one of the foremost riding establishments in the country, approached my father, Colonel V. D. S. Williams, then chairman of the British Horse Society and also chairman and show director of the International Horse Show; he suggested that there was a need for another major London show. As Collings pointed out, coming in July the International, then held at the White City, could never be regarded as the climax of the show season. He felt that there was room for an event at the end of the season—a Champion of Champions show.

Though interested in the idea, the council of the BHS was somewhat luke-warm in its response. They felt, reasonably enough, that to run one big show was gamble enough for a society extremely short of funds; they would never carry the membership if a second show was suggested. They also feared that if the qualification for this show was to be championship status, it was going to be extremely difficult and invidious to define satisfactorily exactly what a champion was and which champions were eligible. Officially, therefore, the idea was dropped.

However, my father found much in the proposal that was attractive and accordingly approached Mike Ansell. When Colonel George Ansell, Mike's father, had been killed in 1914, he had expressed a wish that my father should act in loco parentis. At this time, father was only a subaltern in Colonel Ansell's regiment, so such a responsibility was a remarkable testimony to his ability. Since Mike Ansell was nine years old he had virtually been one of the family, and my father had always been second to none in his admiration of Mike's drive and ability. Indeed, as chairman of the International Horse Show when it was revived at the White City in 1947, he had made Mike a show director entirely responsible for the jumping events.

He was not surprised to find Mike a good deal more enthusiastic about the proposed new show, the Champion of Champions Show, than the more cautious council of the British Horse Society: but as he pointed out, if the show was to be, as suggested, in the autumn or winter, then obviously it would have to be an indoor show. There seemed every justification, therefore, in paying a visit to the popular indoor show in Paris known as 'Le Jumping'.

They were scarcely prepared for what they found. Le Jumping was far more than just a horse show. To begin with there was audience reaction and audience participation such as they had never previously witnessed at any show in the world. The crowd appeared to consist not of the usual horsey, horse-show set, but quite simply of the people of Paris: the sort of people that might be seen at the circus, or the vaudeville, and they appeared to be experiencing the same sort of enjoyment as audiences might get from watching acrobats or dancers.

And how totally involved they became. Colonel Harry Llewellyn, paying his first visit to Paris, with his immortal Foxhunter, having a few weeks previously won the King George V Gold Cup at the White City, received a wildly enthusiastic reception every time he entered the ring. The real favourites, of

(left to right) Harry Llewellyn with Foxhunter, Alan Oliver and Wilf White at the Horse of the Year Show at Harringay in 1951 (*Odhams Press*)

course, were the French riders, the Chevalier d'Orgeix and the effervescent Jonquier d'Oriola, himself a brilliant cabaret artist, together with the irresistible, petite, but outstandingly efficient horsewoman, Michèle Cancre, only just seventeen years old. These were the darlings of the crowd and they were given the kind of reception reserved a decade later for the Beatles.

There was, too, a gala atmosphere about it all. The horse show arena with its gaily coloured fences, flowers and spotlights, seemed more like a theatre or, perhaps, the 'corrida', a Spanish bullring. There was gaiety, enthusiasm, excitement and, understandably, partisanship. Yes, the crowd was immensely partisan, though always fair. Missing nothing they cheered everything worth cheering—and laughed, at whoever it might be, if something amused them. There was also intimacy. The spectators were so close to the riders that they seemed almost part of the show.

Of course, it was recognised that the French were excitable, but they were not normally thought of as horsey, and yet they appeared to be so involved. Clearly, they *were* involved, for a whole week, bringing each performance and the show itself to a level of climax far more associated with a football match—or with the one and only Piaf.

But surely, thought the visitors, if the more-or-less unhorsey French could work themselves up to such enthusiasm over horses jumping, then how much more enthusiastic would the much-more-horsey British be: or would they retain their starchy British reserve, lacking the volatile Gallic temperament? The two visitors to Le Jumping flying back to England did not believe so. Both were convinced, to quote Mike Ansell, that they were onto a winner.

But how was it to be brought about? Who was to put up the money? The BHS had firmly declined to be involved. What about the British Show Jumping Association (BSJA)? Ansell was already chairman and, since his election four years earlier, had left little doubt that what he said went: if only because other members of the BSJA executive, all contributing unique talents, appreciated that they had a man at the helm with the sort of drive and imagination that could lead them to the top. But where to hold a new show?

All agreed that it had to be indoors: outdoors in September or October was too big a risk, but suitable indoor arenas in London were few and far between. The new Earls Court was considered far too big. For various reasons the Empire Pool at Wembley was discarded. Olympia, had it been immediately available, was considered too closely associated with the International Horse Show which had been held there from 1907 until 1939.

Somebody suggested the arena at Harringay which at the time was housing a circus; after all it was owned by the Greyhound Racing Association (GRA), as was the White City. It did not seem too propitious an idea. Situated in the unfashionable north of London, Harringay suggested marts and markets rather than a horse show. True, for many years the Hunter Show had been held at Islington not far away, but to most people Harringay was an unknown and indeed not a wholly desirable district. Nevertheless, it was a mere twenty minutes by underground from Piccadilly and anyway there seemed to be no alternative. A visit was paid to the circus: enjoyable in itself but not particularly prepossessing as a site. The arena was neither really one thing nor another: it was not specifically too big nor too small, but there was no 'pocket', no area where the next few horses could be held before entering the arena. Indeed, there seemed to be no collecting ring at all, and very limited car parking space.

It had, however, two assets. Adjoining it there was an outside stadium which could be used for the preliminary judging of the show classes. More important, it was owned and administered by the GRA with which, under its Chairman, Mr Frank Gentle, such a happy relationship had already been forged at the White City.

The BHS, responsible for the International, as still lukewarm, but the BSJA was now firmly behind the new show though without the financial resources to mount it. Finally it was arranged that the BSJA should stage it, but the GRA, which had already backed the International Horse Show, should be invited to underwrite it. With considerable hesitancy and a certain amount of scepticism they generously agreed, impressed with the calibre of the people responsible for the International, and with their determination and enthusiasm.

Tony Collings was summoned to London and his own enthusiasm matched the others'. Within hours they were getting down to planning. His original idea had been to run a show comprising every sort of class—hunters, hacks, harness, ponies—but to make it only open to horses that had won at the major shows during the season. Le Jumping had been devoted entirely to jumping.

Tony was loath to forego or diminish the importance of the show classes. But Mike Ansell, an instinctive showman, appreciated from the start that it was the jumping that would pull in and excite the crowds. Perhaps it was my father, with his close association with both the BHS and the BSJA, who suggested the compromise: it would be a BSJA jumping show, but there would be a leavening of show classes open only to qualified horses and ponies. This was the pattern that was adopted and it lasts to this day: though, paradoxically, it was the show classes which gave

Captain Tony Collings: he had an idea (*David Frier*)

this great jumping show its title. Collings had always insisted that this was to be a Champion of Champions show: only those horses and ponies which had won prizes at major shows during the season would be eligible: thus in each section there would be a 'horse of the year'. This was how they talked about the show in the planning. This was the title that became accepted: the Horse of the Year Show—the grand finale.

Convinced of the viability of the idea, more than this, carried away by an almost total conviction that they were onto a good thing, discussions and planning continued apace. A committee and working party were formed to consider the whole idea but it was, in reality, simply a formality. The determination to go ahead was invincible. Not for the first time, and certainly not for the last, Mike Ansell had well and truly got the bit between his teeth.

He was supported not only by my father and Tony Collings, but by such people as Colonel Harry Llewellyn, the late Mr Bob Hanson and Mr Ruby Holland-Martin who had the vision to share his enthusiasm and encourage it; and such as Colonel Guy Cubitt, the late Brigadier John Allen and the late Mr A. H. Payne, the owner of horses of the calibre of Red Admiral, ridden for him at that time by Alan Oliver. The BSJA was also represented by stalwarts such as Mr Dungworth and Mr Hindley. Altogether an able, experienced and forward-looking team, though it has to be admitted that all concerned regarded the outcome Show as highly speculative: an indoor show, at Harringay N4, a horse show to be known by the somewhat fanciful title, Horse of the Year!

There were plenty of sceptics in the autumn of 1949 as the third week of September, the date fixed for the show, approached. Fortunately the faith of those most immediately involved never wavered.

# 3 Take-off

Doubts—non-existent advance bookings—
encouraging entries—overseas riders—a good team—
names that were to become famous—old stagers—
irksome delays—Pat Smythe—last-night
enthusiasm—a substantial loss

Their faith, however, was sorely tested. No one was under any delusions. It was not likely, at the beginning, to be a money spinner. Considerable losses had to be anticipated, only too clearly confirmed by the paucity of advance bookings. Although the cover of the programme boasted 'The Horse of the Year Show presented by the British Show Jumping Association' it was perhaps more accurately described on the inside, as *organised* by the BSJA, for certainly the BSJA had no intention of footing the bill.

Once again Frank Gentle of the GRA had to come to the rescue. It was he, in September 1945, just four years earlier, who had been so impressed by the Victory Show Jumping Championships put on at the White City, and organised by Mike Ansell. On the strength of it, he had undertaken, on behalf of the GRA, to stand all the losses for the first three years of the International Horse Show if it were revived at the White City instead of at Olympia where it had been held before the War. As a result the International—it became the Royal International ten years later when it celebrated its fiftieth anniversary—had become firmly established as a show of major importance, and a vital date in the international calendar.

For one year at least the GRA agreed to underwrite the losses of the new show, after which it would be fairly clear whether it was going to prove feasible or not. There were plenty who had their doubts; not surprisingly when it was learned that on the opening day, 13 September—not, perhaps, the most appropriate date for an opening as a cynic was heard to remark—there was precisely £69 in the till.

But the entries were good—30 in the hunters, 31 in the hacks, 15 cobs, 36 ponies, about 125 adult British jumpers, with another 15 from overseas, and the same number of juniors. With the dressage competition—then described as the 'Best Trained Horse'—and the handy hunter competition, there was a total of about 400 entries. This, by any standard, for a first-time Show

Major David Satow: general factotum (*Field*)

was encouraging. The overseas entry from Belgium, France and Ireland was particularly interesting for it included the legendary Chevalier d'Orgeix from France, with the adorable Michèle Cancre (they were later to marry, and later again to separate), the impressive and experienced Captain Chevalier de Selliers de Moranville—a real mouthful for the announcer!—from Brussels, with his already-famed Sea Prince; another popular Belgian, Willie Waterloos, and from Ireland, two horses that were to become part of the Horse of the Year Show history, Hack-on and Go-Lightly.

It was a good team, too, that had been assembled to run the Show. Frank Gentle, in recognition of the responsibility he had undertaken, was nominated joint show director with Colonel Ansell. Captain Jack Webber, for so long secretary of the BSJA, later its chairman, was assistant Show director. Phil Blackmore, one of the BSJA pioneers, was clerk of the course while his son, John, now secretary of the BHS, was in charge of the office. Clerk of the scales was Jack Grayston, whose brother Bob was the leading show jumper in South Africa for so long, and who became known to hundreds as the chief collecting ring steward at the White City. Len Dungworth was chief arena steward (the story that Mr Dungworth was *collecting* ring steward was entirely apocryphal!) while I was invited to be responsible for the public address. With the retirement of Jack Webber in 1973 I am the lone survivor of that team, with the exception, of course, of Mike Ansell, though he himself missed the Show in 1972 and retired as Show director in 1975.

Two other names should be mentioned as they were greatly loved and respected in the early days and sadly missed when they died: 'Doc' Nichol, the medical officer and Mr Steele-Bodger the veterinary surgeon. Secretary of the BSJA who also acted as show secretary at the beginning was W. S. Pearson; Major David Satow, who was to die so tragically in 1970, acted as Mike Ansell's personal assistant and was generally a tower of strength.

The organisation was certainly strong as, in fact, was the list of entries. Any show today that could boast the names which entered that first show would be assured of success, though at the time, most of the riders had hardly been heard of: Pat Smythe, still in her teens and almost completely unknown; Peter Robeson, also in his teens, and young Douglas Bunn just out of them; Harry Llewellyn, whose Foxhunter was already a name to be reckoned with having won his first King George V Gold Cup at the White City in 1948 and, a few weeks earlier, helping to win a bronze medal for Britain at the Wembley Olympics; Ruby Holland Martin, regularly representing Britain until crippled by a serious hunting accident; Lulu Rochford, Bay Lane, Wilf White, whose great Nizefela was just beginning to win the public's admiration; Mary Whitehead and Brian Butler, one of the first civilian British riders to win a major international event.

Among those who had already made reputations for themselves before the war were such people as Bill Clarke, Tom Taylor, Len Carter, Syd Woodhall, the Massarellas, J. W. Creswell, J. Woollam, and Lady Wright, who as Miss Bullows could be described as the original Pat Smythe or Ann Moore, unless that title should be claimed by Stella Pierce, who when Mrs Leonard Carver won the Grand National with ESB in that dramatic Devon Loch race in 1956.

Some of the juniors' names at that first Show make interesting reading too: Pat Moss, Ted Edgar, Jill Palethorpe (later of Earlsrath Rambler fame) Maureen Styles, Yvonne Fossey, George Hobbs, David Barker (described in the programme as Charles D. Barker), Betty Nichol, the doctor's daughter and Alan Oliver. All were to make names for themselves in the future.

Nor should one omit the names of some of those taking part in the show classes: Reg Hindley, who was to captain the British three-day-event at Helsinki; my stepmother, Brenda Williams, the first rider to represent Britain at dressage in the Olympics; Count Robert Orssich, the doyen of showmen; Pamela Carruthers, now course builder at Hickstead—she also did well jumping with Galway Bay which was later ridden by Alan Oliver; Miss Diana Mason, Miss Anne Hammond, Mrs Slane Fleming, Mrs Irene Macintosh, Mrs Sheila Inderwick, Mrs Davina Lee-Smith Wightman, all leading figures now in the field of higher equitation.

There was certainly no dearth of talent at those first performances on 13 September 1949, only a dearth of spectators. True,

Two early leading lights relax: Pat Koechlin Smythe and Alan Oliver (*Roberts*)

thanks to a certain amount of 'paper', the evening performance was quite respectable: and for the first time those involved felt that the Show might succeed. There was an unmistakable atmosphere, a tension which is not usually experienced at a horse show. Could the Horse of the Year Show build up into an English Le Jumping? It seemed *just* possible even that first day, despite the virtually unavoidable teething troubles. These for the most part centred around the building of the courses. Mike Ansell had been a little over-ambitious in attempting to include three jumping events in one performance. As each course took at least twenty minutes to build, and as all the fences had to be trundled in on great trucks loaned from the circus, there were inevitable delays. Between the jumping events there were only the Crawley and Horsham Pony Club activity ride, excellently put on by Jack and Vera Allen, and a dressage display by Monsieur Krier from L'Etrier Belge.

On one occasion there was that sound which to a producer or impressario is the most dreaded of all, the slow handclap, which went steadily on until drowned by a mighty crescendo from the organ. Unable to afford a band, a well known organist from Blackpool had been engaged. Unfortunately his organ was mislaid between Blackpool and Harringay, so arrived too late for a rehearsal. At the first performance it was not only excessively loud, but it went through the whole gamut of appearing and disappearing, emitting a profusion of coloured lights—not perhaps the most dignified accompaniment to a horse show, as the audience made all too clear.

Jumping under the FEI (Federation Equestre Internationale) rules was still somewhat suspect among the diehard members of the BSJA, and so there had to be national as well as international classes. Fortunately the jumping in the international classes quickly developed into a duel between Britain, led by Harry Llewellyn, and France led by Jean d'Orgeix. Llewellyn won the first class from d'Orgeix, then Michèle Cancre pipped them both: and so the Anglo-French duel developed throughout the show. In the ladies' event the Irish Iris Kellett, already winner of the Ladies' Championship at the White City, beat both the French Michèle Cancre and the British Mrs Harry Llewellyn riding Foxhunter. (Iris Kellett recovering from appalling injuries from a fall came back to win her second Ladies' European Championship in 1972 and then produced those great horses, Easter Parade and Pele, for Eddie Macken and later Paul Darragh to ride.)

The final event went to d'Orgeix and Sucre de Pomme, which was very satisfactory from an international point of view. Yet, surprisingly, it was a national event that aroused the most interest, and possibly the most enthusiasm. Appropriately, it was

the Leading Show Jumper of the Year class.

Pat Smythe, as a teenager, had bought for a very small sum of money a little bay mare, Finality, scarcely 15 hh high. Finality had been bred from Kitty, the old mare that used to pull the milk float in the village to which, as a schoolgirl, Pat Smythe had been evacuated. With the remarkable natural skill and dedication with which she so quickly became associated she soon brought the little mare to the top. At the end of 1947 for rather complicated reasons, when for the first time Pat had been invited to represent Britain, Finality was sold to Tommy Makin, who then sold her on to a great supporter of show jumping north of the border, Mr Jimmy Snodgrass from Midlothian.

Pat's financial situation was such that having recently lost both her mother and her father, she was in no position to prevent the sale. Perhaps, too, she was confident that as she had been able to bring Finality so quickly to the top she would soon be able to find other horses to buy cheaply and bring on in the same way (as she very quickly did: Tosca and Prince Hal).

So, sadly, she let Finality go: but during the 1948 season Finality and her new owner never really hit it off. Nor did their performance improve in 1949. Shortly before the Horse of the Year Show, Jimmy Snodgrass 'phoned Pat Smythe and asked her if she would ride the horse for him if he brought it down to Harringay. She agreed, though she had not ridden her old favourite for more than a year. Their unique rapport, however, was quickly re-established. In the Horse and Hound Cup on the first afternoon she finished third from a field of fifty (Peter Robeson on Craven A was second, while the winner was the little known north country girl, Gene Whewell, Alan Oliver's first wife, with Spring Meeting).

The main event of the second day was the Leading Jumper of the Year. The entry was so big that it had to be divided into two sections and a final. But Pat and Finality eventually won the first prize of £200 (there was not even a cup to go with it!) from none other than Ted Williams on a great horse called Tim later ridden by Bay Lane, and still jumping some fifteen years later when the Show had moved to Wembley, to be ridden then by the youthful Paddy MacMahon. Two good horses Niblick and Gay Lady were third and fourth, both owned by Mr J. Woollam from Monmouthsire, a county already boasting the Llewellyns, later to produce the Broomes, the Johnseys and Richard Meade.

Pat's victory on the diminutive Finality brought the crowd to its feet. She became the darling of the Show. To have Michèle Cancre and Pat Smythe, Colonel Harry Llewellyn and Chevalier Jean d'Orgeix all in one show, what more could be asked? If it had done nothing else, the Horse of the Year Show had created new, exciting personalities, moreover, with a new public. The crowds attending Harringay were totally different from the White City crowds—for the most part they were Londoners, for whom show jumping was a new sport.

It certainly aroused the enthusiasm of the crowds, especially on the last night of the three-day show. Virtually a sell-out, thanks to the word getting round that it was a good show, something different, and 'you ought to see that French girl', there was a wonderful atmosphere from the start.

Tommy Makin's very popular Snowstorm won the first event of the evening, just pipping the new favourite of the crowd, Finality; the Hunter Championship was won by Reg Hindley's magnificent Mighty Fine, also champion at the International, from Hugh Sumner's Blarney Stone, a pre-war champion. The team jumping was won by Ireland with 24 faults to France's 28—just one fence down; Britain was right out of it, but the prize for the best individual performance was won by Llewellyn's Foxhunter, a performance that again brought the crowd to its feet.

So on to the Cavalcade, the great assembly of all the winners, and the epilogue which it was my responsibility to compose and recite—a task increasingly difficult year after year until a certain Ronald Duncan came to the rescue: but that was some five years later.

Enthusiasm for the Show was perhaps disproportionate to the attendance—certainly to the takings, the loss amounting to nearly £2500. But the GRA paid up and those involved cautiously but confidently scented victory. Each day the Show had gathered momentum. The last night had been a triumph. Problems there had been, difficulties to overcome, weaknesses to be ironed out, but the fact that the Horse of the Year Show had in it ingredients for success could not be doubted by anyone involved with that first Show.

# 4 Getting Into Its Stride

Solving the teething troubles—the heavy horses—archaic rules—'against the clock'—guarantors—profits—show-class inconsistencies—judging methods—a ducal raspberry—a possible solution—Tony Collings' tragedy

Very soon after the Show closed, in the comparative triumph of the final night, Mike Ansell called a conference which, in fact, was a kind of post-mortem, though the emphasis was as much on improvements in the future as on weaknesses in the past. The main problems, apart from attracting capacity audiences, appeared to be the delays between the jumping events, the method of judging the show classes, and the rules still used for the national classes in the show jumping.

Over the next year or two these problems were ironed out; in the following particular instance, Mike managed to turn something that might well have been accepted as an insoluble obstacle into a positive asset. It was obvious that with so much jumping the arena, consisting principally of tan, needed constant harrowing. Though small compared with an out-door show, the size of the arena meant not only that the harrowing was a lengthy operation, but from an audience's point of view it was a tedious one too.

Before the war, in the early thirties, Mike Ansell had visited the Toronto Winter Fair. He had watched ice hockey and now remembered how, during the intervals, the rink had been swept by an arena party on skates, as a drill and in formation. Could not something similar be done at Harringay? What was normally used to pull a harrow? In modern times, tractors but previously, heavy horses. Why could not harrows pulled by heavy horses be used to harrow the arena? Further, why could they not do it to music, as a kind of musical drive?

Thus was born the famous musical drive of the heavy horses, still one of the most popular features of the Horse of the Year Show. At first they appeared as one of the five sections in a display entitled 'The Horse: Servant and Friend of Man'. At each performance two or three of the sections would give a short display: stallions, light horses, ponies, harness horses, heavy horses. When it was the turn of the heavies, at first they simply paraded round with their harrows. Later, their display was

Whitbread and Company Ltd's Hengist and Horsa, both 18 hh, leading the Musical Drive of the Heavy Horses (*Lane*)

developed into a musical drive with complicated patterns to carefully selected music.

The drive was first fully launched in 1952. Later it was to have music specially chosen by Sir Malcolm Sargent. The climax of the drive is the complicated 'cart-wheel', a remarkable driving achievement in itself; but it is the surge down the arena of twelve of these magnificent animals abreast, to the strains of *A Fine Old English Gentleman*, that brings a lump to the throat.

Originally there were just four teams with two horses in each: Shires, Suffolks, Percheron, Clydesdales. Later this was increased to five with a team of grey Shires joining the black Shires. Finally there were six teams, the Percherons and Clydesdales sadly dropping out to be replaced by more Shires. The ten minutes occupied by the drive has not only given immense pleasure to the audience, which night after night, year after year, gives it a

David Broome on Sportsman
(*Findlay Davidson*)

rapturous reception, but it leaves the arena in perfect condition, harrowed into every corner.

Remarkably, the old BSJA rules were retained for the first five years of the Show, such was the opposition to any changes. Today they make curious reading with '4 faults if a horse or pony causes all or any part of the obstacle to fall with forelegs: 2 faults if a horse or pony causes all or any part of the obstacle to fall with hindlegs. If a horse or pony or rider falls: 4 faults. 1st turn round, circle or refusal 2 faults: 2nd turn round, circle or refusal 4 faults. If a horse or pony jumps wing, whether wing is knocked down or not, 4 faults'.

Slats, those thin white laths that used to be lain along the top of a pole and could be dislodged for half a fault, had fortunately been abandoned shortly after the war, largely because they were always being blown off by the wind; or riders insisted that it was the wind that had blown them off if they were faulted. Fortunately, too, there was no water at Harringay for on this point the BSJA rules were complicated in the extreme. '1 fault shall be added if a horse strikes the take-off fence hard, or knocks it down. Each foot dropped in the water 1 fault. If the horse jumps to the side of the water it shall count as many faults as would have been made had the water been wider'—whatever that may mean!

The international riders found these rules quite unacceptable. Although, as a sop to the old guard of the BSJA, national classes were included in the programme these were reduced to a minimum though to considerable BSJA displeasure. Even in 1951, there were only five BSJA classes to nine international. Before long they were down to just a couple of events.

From an audience's point of view, and therefore from the point of view of presentation, it was of considerable importance. In BSJA rules there were no time faults, and so no time allowance; the time limit of the course was generous: 300 yards per minute; there was no jump-off against the clock—all of which resulted in a somewhat leisurely competition.

Under the international rules the time limit was tighter: 350 to 430 yards per minute; the time allowance was half the time limit; for every second or part of a second over the time allowance there was quarter fault; in the event of equality of faults there was a jump-off against the clock. It was this which was so resented by the BSJA riders. They believed that jumping big fences at speed could be both dangerous and bad for horses. Obviously there was a certain justification for their criticism. Though good course designing and construction should eradicate the danger element there are many who still feel that to ask a horse to jump at speed makes it careless. This is the reason why there is criticism today of the apparent enthusiasm on the continent for

competitions against the clock in the first round.

From the spectators' point of view, however, it cannot be denied that a competition which finishes—as do all competitions now under both rules—against the clock is far more exciting. Indeed a competition which does not run to a jump-off, as occasionally happens, is considered an anti-climax.

To coax the co-operation of some of the older guard of the BSJA in this matter needed all Mike Ansell's tact and determination. Though he achieved it there was inevitably a residue of suspicion. For this reason he felt that to ask the BSJA to back the Show financially was still quite out of the question. The loss of £2400 the first year was reduced to £800 the second year when the length of the Show was increased to four days and the date put back a fortnight.

Once again the GRA had paid up, but Mr Gentle now made it clear that the Show must forthwith stand on its own feet. With the BSJA reluctant to co-operate to any extent financially and the BHS still in no position to guarantee the Show, there were now only private individuals to whom the Show could appeal. Sponsors at that time were unheard of but the response, as so often in these circumstances, was prompt and generous. Some twenty guarantors undertook to write off any losses sustained in 1951 and thereafter. In fact, although the cost of putting on the Show in those days was about £10,000—it is nearer £100,000 today—the Show was very soon making a 20 per cent profit.

At the time of writing, the BSJA, together with certain charities which get the profits from the gala night each year, has benefitted to the tune of well over £300,000 in about 25 years at a minimum risk. On the other hand the BSJA has made a remarkable contribution to the Show through its expertise. The Show has resolved itself, since those early days, into a very happy and successful partnership, embracing the Association itself, its members and the other elements of the horse world that are represented by the British Horse Society, including both show class exhibitors and administration.

Oddly enough, it was the show classes which were another bone of contention in those first years. When Tony Collings had originally suggested a champion of champions show it had never been envisaged that it would consist principally of jumping events. Originally the idea had been to find the supreme champion of the year in the different sections of the horse show world: hunters, hacks, cobs, ponies, harness, and possibly animals shown in-hand. It was essential, therefore, that although the format of the Show had altered somewhat from its original conception, show classes should still be included and, indeed, should play a prominent part.

Eddie Macken on Boomerang (*Findlay Davidson*)

Alwyn Schockemohle on Rex the Robber (*Findlay Davidson*)

Virtue out of necessity: the arena is harrowed by the heavy horses (*Lane*)

There were two problems. The first was that now that the Show was attracting an entirely different type of audience, whose main enjoyment was the excitement of a jumping event, show classes were felt by some to be superfluous. But there were many on the committee who believed strongly that to drop the show classes would be to betray the original idea. It could not be denied that the new public, for the most part, had little knowledge about horses and were not interested at all in the finer points of conformation and movement, action and presence so that for them, show classes were a very protracted and intensely boring affair. As it was pointed out, even at a big county show, even at the International or the White City itself, the show classes were relegated to the early morning before any audience had gathered apart from a handful of aficionados and those involved.

The only solution was to hold these classes elsewhere, only bringing the animals into the arena for the final judging and presentation of awards. Fortunately, there was an outside arena at Harringay used for greyhound and dirt-track racing. It was far from ideal, and this led to the usual grumbles—many of them more than justified—but it was a solution.

So the custom of carrying out all the preliminary judging in an outside arena was introduced. This gave the judges plenty of time, and although understandably, the exhibitors complained that they were only given a few minutes in the indoor arena in front of an audience, it satisfactorily achieved the object of maintaining the tempo of the Show.

It is interesting that today there is a growing feeling that more

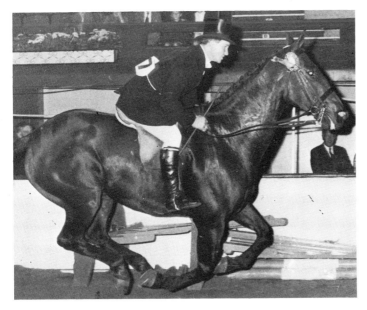

Show hunter, State Visit, ridden by David Tatlow (*Lane*)

time in the indoor arena should be given to the show classes. This is testament to the greater knowledge about horses and ponies now possessed by an increasingly large number of people. It is also testament to the greater influence of those concerned with show classes, thanks to the excellent standard of their exhibits. The championships, in particular, are given a more generous time allowance indoors. There is no doubt that the public enjoy them, not least because the commentators are able to build up the involvement of the crowd by giving it information which makes the show classes more interesting.

No one, today, resents the inclusion of the show classes in what is essentially a jumping show, a show that is generally, though not wholly accurately, assumed to be the BSJA's own Show. The show classes make a welcome respite from the jumping for though there has been ample time for the preliminary judging of the show classes in the outside arena, the final judging inside is allowed very little time indeed. Except for the Championships it really is a case of in and out, which owners of animals that have cost a lot of money to purchase and produce, understandably resent.

The other bone of contention, regarding the show classes was the method of judging. With the intention of finding the champion of the year it had been proposed by Tony Collings that from the beginning of the showing season in May, right through to the Horse of the Year Show at the end of September, a points system should operate whereby the winner of an open class at any show affiliated to the BSJA should collect 3 points if the first prize did not exceed £5, 6 points if it did not exceed £10, 9 points if it did not exceed £15, 12 points if it did not exceed £20 and 15 points for any first prize worth more than £20. For second, third or fourth there were proportionately fewer points.

Points were also awarded at the Horse of the Year Show, but the winner of the title, Hunter or Hack (or any other) of the Year, was not just the winner at the Horse of the Year Show, but was the one that had collected most points through the year. Or was it? The system led to a good deal of argument and controversy. It was possible that two horses, both the winner of the class at the Horse of the Year Show, worth £100 incidentally, and the most consistent horse with the largest number of points collected through the year, could claim to be the Hunter of the Year.

The first year the hack, cob and pony class winners presented no problem, Liberty Light, Knobby and Legend each respectively winning both sections. But the hunter class at Harringay was won by Mr Reg Hindley's Mighty Fine while Norwood Unique, owned by Mr Bob Hanson, that great patron of show jumping, and ridden by his son, Bill, an international show jumper who was to die so tragically a year or two later, was the most consistent. This presented no real problem because both Reg Hindley and Bob Hanson were great sportsmen, and both were dedicated to the success of the new show. It soon became apparent, however, that certain exhibitors, not always with the best horses, were chasing every point they could pick up. There developed the anomalous position of a horse which stood well down the line in the Harringay hunter class winning the title of Hunter of the Year because, during the season, it had collected the most points.

An attempt was made to rectify the anomaly by awarding 100 points to the winner at Harringay, but it was not wholly successful. Anyone who was really prepared to scour the country for points, especially if they had a horse good enough to stand even fourth at Harringay, so picking up some 20 points, could still build up an overall lead.

The matter came to a head in 1952 when one of the hunter judges was the Duke of Beaufort. His co-judge was the Earl of Halifax and quite apart from their knowledge and experience as judges of hunters, they were men of considerable stature who commanded respect from every point of view. Their own selection in the hunter championship did not coincide with the horse which had collected the greatest number of points, and which was therefore, officially, the Hunter of the Year. At the presentation it was the custom to line up the awards of the class, hand out their rosettes and then call forward the winner of the title which could, of course, be a horse not even in the line-up. On

'Tub' Ivens with 1974 Lloyds Bank In-Hand Champion, Sammy Dasher (*Davidson*)

this occasion the title winner stood about fifth in the class.

As he came out of the arena the Duke of Beaufort turned to me at my commentary desk which was right by the exit, and upbraided me in no uncertain manner. 'You've made me look a complete fool', he said. 'That wasn't the one I wanted to win at all.' I saw his point completely; indeed, I had been opposed to the system from the beginning as it was so obviously open to blatant abuse. But it was hardly my fault!

Fortunately His Grace lodged his complaint in the right quarters, supported by his distinguished partner, and this quickly achieved results; but the old system was to be replaced by a method of judging which over the years has been not a lot less controversial. This new system of judging really amounted to judging the class in two sections with one judge to be responsible for conformation and action, and another judge to be responsible for the ride. Each could award up to 40 points, the two judges then coming together to award up to 20 points for general presence.

In theory this was a good idea because it prevented judges being influenced by form: the difficulty was that the two judges might use entirely different yardsticks in their marking. One

Final judging of a show class (*Stonex*)

judge might give his best 40 points and his worst 25 points. The other might give his best 25 points and his worst 5 points, which would mean that one judge's influence was much greater than the other's.

It was hoped that this discrepancy could be rectified by the 20 marks at the judges' disposal when they came together at the end on, ostensibly, the general presence. It was assumed that with these few marks to play with they could adjust any very obvious anomalies, such as finding a very ugly horse finishing up with the best marks, or a horse that was a dream of a ride standing at the bottom. Up to a point this worked; in some cases, too well, because it was not unknown for judges to mark all the exhibits so uniformly that there was really nothing between them, the positions being decided entirely on that final 20 points.

I have to admit that this system has produced a number of not very satisfactory results and that it has been frequently abused. Nevertheless as one who strongly resents judging on form, I believe there is still a lot to be said for it.

My own idea which has been considered but never adopted, perhaps for very good reasons, is for each judge to judge the class independently and place it. If there were 20 in the class the horse at the top would get 20 marks, the second 19, the third 18 and so on. Alternatively, one could do it the other way round: the horse at the top would get 1 mark, the second 2 and so on. Or it may be found better in practice to have a differential of 2, the horse at the top getting 40, the second 38 and so on.

Assuming that the conformation and the ride were of equal importance—though the comparative value could well be adjusted if it was considered that conformation was more important than the ride—the two relevant scores would have been achieved by each exhibit: these two would be added together. The judges would then get together and between them add up to another ten points for presence; or, if they thought preferable, the judges would only come together in the event of equality of scores.

This would eliminate the problem of any disparity in the marking, but it would still achieve the independence of the judging. Even if one judge did place according to form, his order could be balanced by the marking of the other judge.

Competitors have never liked the existing system, feeling that it enables a comparatively unknown, or unsuccessful horse or pony to carry off the supreme title—Hack of the Year, or Pony of the Year—while a great winner throughout the season could go down just because one judge does not, for some reason or another, like something about it. Admittedly this is a problem, but there must be a certain merit in bringing to the judging of a Show which is really the climax of the whole season, a complete independence, influenced as little as possible by the form book. Perhaps one day the system of judging will be revised or abandoned as, after the Duke of Beaufort's complaint, was the points system.

In 1954 the title of the Hunter of the Year went officially to the winner of the actual class at Harringay. Ironically a few weeks after the decision had been taken to abandon the points system which had originally been suggested by Tony Collings, he was tragically killed in the first Comet disaster. A man of immense charm and ability he had in a few years, built up the Porlock Vale Riding School into an establishment of international reputation. He was a brilliant horseman winning the second Badminton Three-Day Event ever held, having been second in the first. The original inspiration behind the Show, he had been an industrious and live-wire member of the committee despite the distance he lived from London. His death was the first of disasters that were to batter Mike Ansell as director of the show: disasters that invariably seemed to make him even more determined to succeed.

# 5 The Personalities: a Virtue out of Necessity

Need for variety—early displays—personalities—High and Mighty—signature tunes—Grand National winners—Arkle, Mandarin, Hyperion—a virtue out of necessity

It had become obvious early on that it was essential to have as much variety as possible in the programme. The long pauses while the jumping courses were being built were both unprofessional and created unwanted delays in the performance. It was also very soon appreciated that non-stop jumping quickly became boring to an audience. A leavening of show classes was not enough at a show which was quite different in conception from the International Horse Show. More entertainment was required, either in the way of displays or parades.

In 1950, to mark his retirement, that great old show jumper, Silver Mint which five years earlier, at the age of 27, had won the BSJA high jump event, was paraded at each performance. It went down extremely well, and as a result there was a much more ambitious parade the following year, under the general title of 'The Horse, Servant and Friend of Man', as described in the previous chapter.

Particularly popular in these parades were horses which could be described as celebrities, horses which were already well known to the public, such as Nickel Coin the winner of the Grand National, the first mare to have won for many years; or Winston the police horse, ridden by the Queen at the Trooping of the Colour. There was the great hackney mare, Holywell Florette, indisputably the greatest hackney of her generation, and the sensational little hackney pony, Bossy, only 11.3 hh but Pony Champion in 1949, 1950 and 1951. Interestingly there was, too, the splendid Shire champion, Leckhampstead Surprise. In 1974 when for the first time the Show included a Shire Horse of the Year Championship it was won by Lillingstone Again—the two prefixes Leckhampstead and Lillingstone being next-door villages in North Bucks, for so long a noted area for Shires.

In 1952 the different sections gave way to a single display, known for the first time as the Horse Personalities Parade. They were led in again by Winston, the police horse, thus starting a tradition whereby the police Horse of the Year leads in and 'controls' the parade, ostensibly guiding each personality to its

Dramatic start to the Parade of Personalities (*Stonex*)

place and directing them all out at the end. In the parade that first year were Teal, the Grand National winner, Emily Little, the Badminton winner, Pretty Polly, the outstanding show pony, the 32-year-old Rosina Copper, an Argentine Polo pony of international repute and, for the first time, the pit ponies. The latter made a tremendous impression. As they were led in to the strains of *Blaydon races*, with the lights lowered and the lamps on the pit helmets of their leaders beaming, we all found it extremely moving.

Harringay memory · the annual appearance of the pit ponies (*Roberts*)

Very quickly, with improved production, the parade became a most attractive item in the programme. At the same time it served a useful purpose, for while the personalities were parading down the centre of the arena in spotlights, the houselights being kept at a discreet level, the fences for the next jumping competition were quietly being erected down the sides of the arena. Thus it was ensured that there would be the minimum of delay before the next jumping started.

Today, the parade of personalities is always the final item before the interval, and it is probably true to say that the interval, scheduled to last 20 minutes, would in any case allow enough time for the course for the main event, which takes up the whole of the second half of the programme, to be built. However, by starting building in the semi-darkness during the parade it does mean that if the programme is running late, the interval can be shortened and the course can still be ready on time. It has to be remembered too, that time must be allowed for the competitors to walk the course.

There are those who feel that the course building distracts from the parade. This could be true for those who sit at the ringside: but these for the most part are officials, owners and riders. I doubt if those higher up the stand are unduly disturbed, though possibly the personalities themselves might be. It is more likely that when the spotlights hit the Trumpeters of the Household Cavalry who have moved quietly into position in the darkness, the concentration of the audience is held entirely by what they can see in the lights. The trumpeters certainly make an effective and dramatic opening to the parade, more so than ever now that they usually start with the super *Fehrbelliner Reitermarsch*, the trumpets being echoed and answered by the full band in its gallery above the entrance.

In 1953 the state trumpeters were strengthened by the presence of a team from the Kings Troop, the Royal Horse Artillery and the mascot of the Argyll and Sutherland Highlanders, a minute Shetland called Cruachan, which is the name of the highest mountain in Argyllshire. (For some years there was a not inconsiderable military influence in the Horse of the Year Show, a

45

relic, probably, of Olympia.) But that year it was again the pit ponies which stole the parade.

Evidence that the parade was not yet an indispensable part of the tradition of the Show lies in the fact that in 1954 it did not take place. There was instead a Pony Club display entitled 'The Cat and Custard Pot', of which more in a later chapter. In 1955, however, the personalities not only returned, but were stronger than ever, at least in numbers. There were no less than 18 groups, although only 15 minutes were allowed for the whole parade. As producer and commentator I certainly had problems. After his experience the first year, Mike Ansell aimed at everything going exactly to time. The result, of course, was that none of the personalities had a proper run for their money, while audiences felt that they were denied the opportunity of getting more than a brief glimpse of such famous horses as Gay Donald, winner of the Cheltenham Gold Cup; Kilbarry, winner at Badminton and Individual Bronze Medallist at Stockholm and, by contrast, a splendid old barge horse called Daisy. In subsequent years numbers in the parade have been reduced. Nowadays there are never more than ten or eleven so there is more time to go round, to everyone's advantage.

Final appearance of Nizefela and Wilf White in the Parade of Personalities 1973 (*Lane*)

'It takes all sorts' Pat Taaffe on Arkle, perhaps the greatest steeple-chaser of all time, and Pearly King Mr B. E. George and Blinkers, bought for £9 at Barnet Fair (*Lane*)

Outstanding Hackney Pony, Highstone Nicholas, led here by Mr J. Jones (*Stonex*)

1956 was a vintage year as far as the personalities were concerned, perhaps the best parade there has ever been. In addition to Black Magic of Nork, certainly the most famous hackney stallion of all time, Pretty Polly who could claim a similar show pony reputation, Miss Gladys Yule's two celebrated Arab stallions, Count Dorsaz and Count Orlando, there were also the recently retired Foxhunter, Colonel Harry Llewellyn's most popular-ever show jumper, and both the Queen Mother's Devon Loch and Mrs Carver's ESB, which a few month's earlier had been associated in one of the most sensational Grand Nationals of all time.

There was also Sheila Willcox and High and Mighty, whose two minutes dressage display was, in the opinion of many, perfection, something unequalled at the Horse of the Show, until Jennie Loriston Clarke's recent displays on Kadett. One evening Sheila Willcox's display was televised and I have been convinced ever since that superb dressage could very easily become popular with the ordinary viewer even though it is always assumed that he only wants to watch jumping. The exquisite harmony, the poetry of motion, surely pleases as much as a perfect performance of ballet.

The music for the personality parade had always been carefully chosen, culminating for the final exit in the well known *My Hero* march, which it was always hoped would reach its final chord as the mounted policeman, always the last to leave the arena, turned at the exit and saluted. Successive bandmasters, or directors of music, have attempted to draw out the music, or speed it up, in order to get this impressive effect with varying success. At the end of the music, coinciding with the end of the parade there is a second's pause before the Trumpeters of the Household Cavalry sound the final fanfare. Then, dramatic black out. Originally background music was played quietly as an accompaniment to the parade, the band going into *My Hero* as the horses began to leave the arena. Shortly before the end of the Harringay days however, it was suggested that each individual or group should have its own 'signature tune', as did the pit ponies. To begin with these were fairly obvious: *Pretty Polly Perkins* for Pretty Polly; *A Hunting We Will Go* for the outstanding small hunter champion, Burrough Hills; *Tit Willow* for Tit Willow, a Pony Club pony; *Waltzing Matilda* for the Australian three-day event gold medallist, Salad days.

When, however, a little stallion presented to the Queen by Messrs Bulgarin and Kruschev on their first visit to Britain was included in the parade, it seemed to me that here was an opportunity to be a little more subtle. After discussing the matter with Major Dean, director of music of the Band of the Royal Army

Walking the course (*Findlay Davidson*)

Service Corps as it was then, we decided on *Stranger in Paradise*. This caused considerable comment and while, inevitably, a few thought it to be in dubious taste, even bordering on lèse majesté, most people were amused. So each succeeding year an attempt was made to find appropriate music which was on occasions obvious, but when possible, more subtle. While in 1961, *The Runaway Train* was probably straight-forward for Tom, the 20-year-old horse which spent its life shunting horses at Newmarket, *I Can't Give You Anything But Love* generally raised a laugh for a huge Shire mare with her well grown foal; as did *I Can't Say No* for a mare, Promise, with her progeny, yearling foal, and a 2-year-old, paraded as a group.

*Strolling* was obvious for Stroller; *The Galloping Major* for Major Guy Cunard and his prolific point-to-pointer Puddle Jumper, and the *Magpie* TV signature tune for the pony Puff. More interesting were *I Did It My Way* for the mighty kick-back Vibart; *There is Nothing Like A Dame* for Meriel Tufnell and Scorched Earth, the first Ladies' Race winner; *If I Had a Hammer* for a polo pony; *Knees Up Mother Brown* for a great hackney pony, Highstone Nicholas, who for his second appearance in the parade of personalities had *Anything You Can Do I Can Do Better*. Quite amusing, too, were *The Night They Invented Champagne* for the beautiful Palomino stallion, Bubbly; *Leaps and Bounds* for the Grand National winner Specify; *London Pride* for the superbly turned out equipage from Rothmans with the two perfectly matched horses, Pell and Mell. Particularly appropriate was *Little Man You Have Had A Busy Day* for a Pony Club pony, Japhet, which for eight consecutive years had been in the West Norfolk Prince Philip Mounted Games team.

As the Show becomes older, nostalgia plays an increasing part for many, and the greatest applause has been reserved for old show jumping favourites appearing from retirement: Tosca, Vibart, Mister Softee, Nizefela with *Boys of the Old Brigade* and *I'm an Englishman;* Craven A with *Smoke Gets in Your Eyes;* for Pegasus, *With a Little Bit of Luck*, and for Merely a Monarch, *If I Ruled the World*.

In a quarter of a century of personalities the long role of honour is memorable and distinguished, a feature of the Show which many look forward to year after year. Indeed there was much criticism when it was decided in 1963 to devote the whole parade to winners of the Grand National, though in my own opinion it was one of the most interesting and impressive parades of all. There were eleven post-war Grand National winners in all, including the 1949 100-1 winner, Russian Hero; Nickel Coin, one of only two mares to win since the war, who won in 1951 when 13 fell at the first fence, only 2 actually completing the course

Anne Muir, a popular figure in the driving events (*Ian Ball*)

without falling; Nicholas Silver who won in 1961, one of only two greys ever to win the National; and the magnificent looking Freebooter, winner in 1950.

It was a fascinating and unforgettable display, though in the opinion of many—all except racing enthusiasts, perhaps—it lacked variety: so in 1964 the original personalities parade was resumed with the Grand National winner, Team Spirit, included, but pride of place going to old Nizefela, who was ridden in each evening by Wilf White.

Probably most people would agree that three horses stand out from all the personalities that have paraded at Harringay and Wembley—Arkle, Mandarin and Hyperion. But the last named, although the greatest of all British-bred Thoroughbred stallions, which should, of course, have dominated them all, cannot be awarded the supreme accolade. Even at his first appearance he seemed to think, as he eyed Tosca, that the arena was his mating box, and behaved accordingly. The moment evidence of his state of mind became obvious orders were given for the lights to be lowered and he was quietly led from the arena. Presumably, people in those days were more prudish!

Mandarin appeared in 1963, the year he won the Hennessy Gold Cup (he was owned by Mme Hennessy), the Cheltenham Gold Cup and the Grand Steeplechase de Paris, the race in which Fred Winter piloted him home without bit or bridle. Also in that parade was the Grand National winner, Kilmore, which Fred Winter had also ridden, and Wyndburgh which had been second to Kilmore, and second on two other occasions.

Fred Winter and his trainer, Fulke Walwyn, were very worried as to how Mandarin would behave with all the other horses, the spotlights, the applause and the building of the fences at the side of the arena. At the horse's first appearance Fred rode him round very quietly at the walk. By the second performance, despite his reputation for being skittish, he was being cantered round, almost on a loose rein. Best of all when he had the honour of being the centrepiece of the grand finale, the cavalcade, on the last night of the Show, he stood there like a statue, his head up, his ears pricked, Winter motionless on his back. It was a very moving performance which none who saw it will ever forget: a tribute, too, to his rider's horsemanship.

Arkle, in his own way, was no less great. Without doubt he was the most popular steeplechaser of his time, and when each evening his owner, Anne, Duchess of Westminster, went out to him in the centre of the arena to give him a lump of sugar, the applause of the crowd all but brought down the stadium. He too was a noble, generous horse to look at. The famous yellow and black colours worn by Pat Taafe brought each performance a

Highlight in 1969: display by the Spanish Riding School of Vienna (*Peter Roberts*)

Cynthia Haydon as Lady Scatterdash in the 1964 'Jorrocks rides again' (*Lane*)

thrill to the audience, which appreciated that they were being privileged to look on one of the all time greats. What better signature tune could he be given than *There'll Never Be Another You*.

Dick Francis on Crudwell, winner of over fifty races in National Hunt racing, appearing in the 1960 Parade of Personalities (*Lane*)

As with the heavy horses and their musical drive, so the parade of personalities had become a virtue out of necessity. Originally brought in as a stop gap at best; more realistically employed to enable the course builders to get on with the building of their fences so that any long delays could be avoided, the parade of personalities has become one of the main features of the Show, and the removal of it would now be bitterly resented by the vast majority of those attending each year.

Mr Joe O'Brien's Black Prince which appeared at the Horse of the Year Show in 1971, and was suitably dressed for the occasion, having won the RSPCA Medallion of Merit for his work for charity (*Roberts*)

# 6 Behind the Scenes

The Show office—the Wembley staff—the stable area—security, or lack of it—the exhibitors' corner—declarations—the caravan park—the outside arena—senior stewards—control—veterinary and medical officers, farriers and caterers—the collecting ring—the unsung heroes

It is the displays and parades as much as anything that fully stretch the necessarily elaborate and complex organisation of the Show. As far as the jumping and the show classes are concerned it is comparatively straightforward: those taking part enter, arrive, compete, collect their prize money and depart—an oversimplification of course but with an element of truth behind it. The displays, by comparison, need a great deal of thought and preparation: not only do the participants have to be selected and invited, but presentation and timing have to be agreed, all of which involves considerable correspondence and telephoning before ever they are finally engaged.

At the Show offices at Belgrave Square there is a full-time staff of five which is also responsible for the organisation of the Royal International Show. Colonel Ansell, whose home is in Devon where he runs a flower market-garden, had his own small but fully equipped office there. In the main room are desks for secretarial staff and two glass-partitioned offices for the assistant Show director, John Stevens, who takes over as Show director from Colonel Ansell in 1976, and the Show secretary.

Until 1964, the secretary for both shows was John Blackmore who then became secretary of the BHS. He was succeeded by Major George Worboys a charming, efficient person who rose from the ranks to be commissioned in his own regiment, the Rifle Brigade, and who brought to his job with the Shows not only an efficient, organising mind, but also a delightful, sympathetic temperament which staff and exhibitors alike found extremely helpful. His death, some six weeks before the 1975 Royal International, was a great blow: but such was the organisation by now established that the arrangements were able to go ahead regardless.

The Show office is in action the whole year round, and it is during the quieter months from December to March that the preliminary planning, which includes the displays, occupies most of the senior executives' time. Over the years a pattern of organisation has emerged, a pattern now accepted as being essential to run the Show effectively and efficiently. The several departments are autonomous, up to a point. The purely secretarial section is presided over by the director at Belgrave Square, until a week before the Show when the whole organisation moves into the rather cramped office accommodation at Wembley. The office staff is the only paid personnel of the whole Show, and considering the size and the complexity of the overall staff this is both surprising and rewarding.

The Wembley staff operates under the general manager, for so long John Connell who, on his untimely death, was succeeded by George Stanton, succeeded in turn, on his retirement by David Griffith, all wonderful friends to the Show. The Wembley box office is independently responsible for all ticket sales.

The stable area, obviously of vital importance in a horse show, covers some two acres and consists of more than 300 boxes which are erected just for the period of the Show. These are adequate and well constructed though until recent times, as the site was on a slope, they were not ideal. Over the years there have been problems both with drainage and on occasions with flooding, problems which now seem to have been solved. The stable manager Mr Leonard Brunt, who succeeded a well known forage merchant, Mr L. F. Jollye, does a remarkable job providing bedding and forage, maintaining adequate water supplies and working efficiently the jigsaw puzzle of filling the boxes almost nightly with different horses and ponies. Frequently there are grumbles at the cost of hiring a box for the night or for the whole

Waiting in the wings, (left to right), Warren Wofford, Harvey Smith, Ted Williams, Caroline Bradley and David Broome (*Roberts*)

week, grumbles which fall on the stable manager; but, in fact, such is the cost of erecting temporary stabling that the boxes are generously subsidised.

There is a great atmosphere in the stables which cannot, perhaps, be imagined by those who just visit the Show as a spectator. It is well worth spending an hour or so in this area to become aware not only of the near twenty-four hour activity in the stabling of 300 horses and ponies, but also of the comradeship of people looking after horses. There is first a feverish excitement as zero hour approaches and the horses are being got ready—'Where's that martingale?' 'Give me the dandy brush!' 'Anyone seen the sponge?' 'Don't forget the blacking.' 'They're calling us already!' Then, when it's all over, the relaxation, the post-mortems, the contentment, the chatting with new-found friends, comparing notes, not only between one girl groom and another—'Of course, the food's awful!' 'We always use stud nuts.' 'What time do you start?' 'We have to live in a caravan.' '*He's* rather dishy!'—but between one nationality and another, for there can be at least six different countries competing at the Horse of the Year Show, and each overseas rider brings his own grooms.

Late at night, or early in the morning, there is the packing up and departing, the farewells and the promises to keep in touch. And all the time there is the police security: not only do the animals, all of which are very valuable, have to be protected from those who might interfere; not only does the tack, also very valuable, have to be protected from theft, but it is even necessary to ensure that horses do not escape.

On one occasion, either because stable doors were not being properly fixed, or thanks to a mischief maker, two horses belonging to one of the Italian riders, Vittorio Orlandi, escaped and were discovered galloping along the North Circular: they could easily have been killed. Mike Ansell, ever a firm believer in attack being the best method of defence, calmly enquired why the groom had left the stable doors open, but not surprisingly, after some fairly sharp exchanges, this led to a tightening up of the stable security arrangements.

The stable manager certainly has a responsible and arduous job. In the opinion of some of the riders one of Mr Brunt's predecessors was considered a little too conscientious and was unceremoniously dumped in the water trough. Senior stewards certainly face a risk with high spirited riders as the excitement mounts during the week. There is, in any case, very much an 'end of term' feeling about the Horse of the Year Show. It is the end of a long season and even though there is so much indoor jumping during the winter these days, there is still an awareness that, for many of the show jumpers and for almost all the show

A show exhibitor leaves the temporary stabling for the arena (*Roberts*)

class exhibitors, from now until the spring everyone is going his different way. For a time at any rate it is goodbye.

There is no doubt that it is because of the inevitable end of term feeling of the Show which is the climax of the season, that there is such a desire among exhibitors to qualify to perform at Wembley. Show class entrants must have qualified at certain specified shows. In the jumping classes riders and mounts have to have won more than £250—a figure which is frequently being raised as prize money gets bigger and there is greater pressure among competitors to get to Wembley.

Exhibitors, of course, can watch the Show when they are not competing from the special exhibitors' seats which are situated near to the arena. One of the stewards with the least enviable job is in charge of these blocks. Every trick is tried to get into this area by people who have not qualified, and are not therefore entitled to be there. Each exhibitor, in fact, is issued with a bracelet stamped and sealed onto his wrist, but there is no lack of ingenuity in finding ways and means of getting round this precaution especially on the great last Saturday night which can well be likened to the last night of the Proms.

Fortunately, for many years the steward in charge here has been Gerald Barnes, greatly respected by the show jumpers and, indeed, all exhibitors. With his wife he has frequently accompanied teams abroad, and having produced no less than three international riders, daughters Mary and Sheila and son Tom, they know the whole scene as well as anyone in the business. Mrs Barnes also has the somewhat thankless task of being in charge of the 'declarations'. Each competitor has to declare if he intends to ride two or three hours before the actual event. If he fails to declare he is disqualified. A rider is empowered to depute someone to declare for him: in the case of international riders, it is usually the practice of the chef d'équippe, the team manager, to undertake the declarations. Obviously this rule has to be strictly adhered to, though every effort is made, especially by someone as knowledgeable and sympathetic as Sylvia Barnes, to see that no one misses out.

Some years ago, Marion Mould at the height of her fame, neglected to declare for a vitally important competition. Every effort was made to find her, but to no avail, presenting a particularly difficult problem for the Show organisation. The public was obviously going to be disappointed if Marion did not compete, but once allowances are made rules are quickly ignored. Every effort was made to discover mitigating circumstances. Indeed, one of the particular contributions Mike Ansell has made to show jumping over the years is that while he can enforce rules without there being resentment, he is also a man with the stature to bend rules when it is in the interest of the sport.

Another steward who has a not always enviable job is in charge of the caravan park. This is a little town of its own for with the increasing cost of hotel accommodation and the understandable desire of owners and riders to be near their horses, more and more people like to live in caravans when they go to a show. There is, however, limited accommodation for 150 caravans in the caravan park, and with stricter regulations particularly with respect to fire, the premium on space gets tighter each year. Nevertheless, those fortunate enough to get space for their caravans make a great party of it with all-night jollifications, practical jokes, high jinks, and inevitably a few complaints.

Over the years there have been one or two unpleasant incidents: the aggressive behaviour of Ted Edgar on one occasion led to his suspension, but a measure of his realistic and professional attitude to the sport lies in the fact that he accepted it with grace, acknowledged his folly, and came back again nearer to the top of the tree than when he left it. For the most part the only problems have been caused by that end of term atmosphere again. Though this might surprise many, alcohol does not play much of a part, for

A vital part of the behind-the-scenes activities: the show farrier (*Roberts*)

show jumpers are largely, if not wholly, teetotal. Indeed, they could never live their seven-day-a-week working lives, or keep up the standards of precision and enormous fitness which are demanded by show jumping today if they were excessive or even more than very moderate drinkers.

The caravan park is all part of the surrounds of the big Olympic stadium and the Empire Pool where the main events of the Horse of the Year Show take place. Apart from the stabling, the remainder of the area consists of the outside collecting ring, where the jumpers have their practice jump and do their warming up, and the outside ring, where much of the preliminary judging takes place.

The steward in charge of the outside ring has a somewhat thankless task being responsible for organising the various classes that occupy the ring from as early as 8.00 am until half-way through the afternoon. His work is made the more arduous since it is up to him to ensure that a class is completed in time to go into the indoor arena for the final judging. This is difficult for two reasons. First the method of judging at the Horse of the Year, described earlier, necessitates that no matter how big the class, every animal from the best to the worst, should have the chance of being properly judged, and where necessary ridden. Secondly, judges do not like being hurried. Immense tact, therefore, is necessary, as well as firmness, to keep to the schedule. More often than not it is a race against the clock.

Sympathy is also required. At the end of judging which may have continued for the best part of two hours, when the marks have been totted up, the short list of just ten fortunate exhibits selected to go inside for the final judging is read out. Inevitably there are many bitterly disappointed faces and even, from the children—only the children?—a few tears, as it is realised that the work of a whole season has come to naught.

The same steward, with his two or three assistants, is also responsible for the classes when they come inside for the final judging. The final judgings are spread through the afternoon performance and on two or three occasions, in the evening as well, so it will be appreciated that the show class steward has a long and responsible day. For more than twenty years Colonel 'Handy' Hurrell had the job. Himself a distinguished horseman, he was at one time chief instructor at St George's, the first national school, which was the forerunner of the National Equestrian Centre at Stoneleigh. Until recently, Lord Lieutenant of Cambridgeshire, Colonel Hurrell had the ideal temperament for the job of Chief Show Class Steward and was extremely popular with both judges and exhibitors.

In charge of the indoor arena throughout the morning session and during all jumping classes is a senior representative of the BSJA. For many years this was Captain Jack Webber, Secretary-General of the BSJA, himself. On his retirement he was succeeded by David Bourne, who must be one of the most experienced stewards in the country, having been leading steward at many of the major shows. It is his duty, again, to see that the Show is running to time and to iron out any problems that may arise during a class, be it a competitor missing his turn, an accident, or a competitor arriving late to walk the course. His eye must be constantly on his watch—at the Horse of the Year Show timing is so exact that the programme for a whole day can be completely thrown out if one class runs late. Owing to the high number of entries, it is even necessary on occasions to continue jumping right through the lunch interval and between the afternoon and evening sessions. Sometimes, therefore, it is necessary to 'bring in a standard', which means asking those who collect more than a certain number of faults to retire. It is up to the steward in charge to have worked out how long a competition is going to take—if 5 horses have jumped in 12 minutes, a class with 35 entries will take 84 minutes: another 6 minutes for 'presentation': total, 90 minutes—and inform the judges.

It then has to be decided whether to bring in a standard or try to speed up the competition. This can be done by getting the next horse in quicker, by announcing the result quicker, by shortening the announcement of each horse. The decision is taken principally by 'control', which has overall responsibility for the running of the Show. This, in fact, has been my department since 1949, with various assistants, Raymond Brooks-Ward becoming my regular Number 2 in 1960 and in recent years, incidentally, doing far more than I do. Control is responsible not only for all the announcing and commentating, but also for the actual running of the Show during a performance. It is from control that the orders go out to send in the next class or competitor; it is control that organises the lighting, when and how to use the spotlights, the electric score board and the information screen; control organises the band, telling it when to play and when not to play; it is control that 'handles' the arrivals of VIPs—'Ladies and Gentlemen, we now welcome to the Horse of the Year Show . . .'

Obviously control has to be manned from the very beginning until the very end of the day—'And that, ladies and gentlemen, concludes this evening's performance'. In addition to myself, the team consists of Christopher Hall, a Tunbridge Wells Solicitor, ex-5th Royal Inniskilling Dragoon Guard and well known point-to-point rider, and Tom Hudson, himself a television commentator and secretary of the Belvoir Hunt.

With our fifteen-hour day we have certainly contributed some mileage over the years from our control point position, which has both its advantages and its disadvantages. We operate from the ringside half-way down one side of the arena. At the back of us is a gangway running the length of the arena, immediately below the balcony, and behind us are steps leading up to alleyway that surrounds the arena. It is certainly the hub of the universe: immediately on our right sit the girls who take out the rosettes—a distraction in itself! Behind them are the doctors and veterinary surgeons. Until a few years ago immediately to our left was the Royal Box, now moved to the end of the arena.

The principal advantage of control's situation, at least from Mike Ansell's point of view, has been that it made it easy for him to come and stand immediately behind us, thus keeping his finger on the pulse of the Show, knowing that he was right at the heart of the whole system. As we also do the television coverage from this point, being able to hear our commentaries helped him to know exactly what was going on without having to depend on the loudspeaker system, or being told. If anything went wrong he was in the ideal place to remonstrate. He missed nothing and from such a position he himself felt in complete control of the situation—as indeed was the case.

From a commentator's point of view, it could be distracting to have so much going on immediately behind. Not all of it directly concerned with oneself. It was not uncommon, for instance, for Mike Ansell to ask a question or point something out while I was actually broadcasting. There is talk of moving the commentary position now that Mike has retired as Show director, though a change of position would have one serious disadvantage for us: our proximity to the stewards' private bar has, over the years, been much appreciated!

Apart from the doctors, veterinary surgeons and farriers, all of whom have a vital role to play, there are two other major departments still to be mentioned—catering and the collecting ring. The former, though working in the closest co-operation with the Show, is the responsibility of Wembley Stadium: nevertheless it is the requirements of the Show which have to be fulfilled, and these are of extreme importance—human nature being what it is, workers who are well fed are much more effective than those who feel that they have cause to grumble about the food. It is a long day for the caterers. Early workers need breakfast; sandwiches and snacks must be available in the VIP room and the stewards' bar (officially known as the conference room), at the end of the Show, some fifteen or sixteen hours later. Stewards using the Silver Mint Restaurant where tickets entitle them to free meals expect to be in and out in the minimum possible time; there can be no delay in the serving of the meal, their job is waiting for them.

Even the party from the royal box expects to have its supper in the twenty-minute interval. Quick service is also expected at the snack bars and the bars all round the stadium: Horse of the Year Show audiences want to miss as little as possible, even if they've seen it all before. Only in the Arena Restaurant is the service more leisurely. Here those fortunate enough to be at one of the forty or so tables overlooking the arena can enjoy an excellent four-course meal lasting throughout the performance, lunch or dinner. The cost of a table for four including the meal, but not drinks, is £25. Most of the tables are occupied by those who have sponsored the major events and the original guarantors. For many years Messrs Letherby and Christopher have been responsible for catering at Wembley, many of their staff, under Colonel Livingstone Learmouth, becoming firm friends of the horse show personnel. Indeed, one faithful waitress, 'Mac' (Mrs McKenna) was given a special present in 1973 to mark her long service with the Show. All the catering staff are very much part of the team.

Finally, there is the collecting ring—certainly a department as important as any in the Show. There are two collecting rings, an outside one and an inner one. The inner one acts as a 'pocket', where the class or display or the next half dozen or so jumpers are held awaiting their moment of entry. The collecting ring steward operates from a box at the side of the entrance just inside the arena. He has telephone connections with control, the outside collecting ring and many other points. He also has a microphone to broadcast to the outside collecting ring and stables and its his job to call down to the inner collecting ring those who are next in line.

One of his assistants is based in the outside collecting ring to ensure the steady flow of those required in the stadium. A hitch of some sort can completely throw out the fluent production of the Show. The collecting ring steward, therefore, has an extremely responsible job. He is the link between control and three or four hundred horses and riders. 'Send in so-and-so,' someone from control barks down the 'phone. It is up to the collecting ring steward to see that so-and-so is ready. 'Curtains!' and one of the soldiers on duty at the entrance draws the curtains and, exactly on time, on comes the next horse or item.

If there is a delay or a hold up, then a lot of people want to know why. Without absolute dependability and efficiency at the stadium entrance, things would very quickly go to pieces. Nothing is worse, in the presentation of a show, than the hiatus of an empty arena.

For twenty years Barry Pride, a Kent estate agent and auctioneer, held the onerous position of collecting ring steward. In recent years he has succeeded Colonel 'Handy' Hurrell as chief ring steward, his brother Roger taking his position at the control box at the entrance.

Only a handful of senior stewards have been mentioned: each, in fact, has his own team of three or four men, making a full total of some forty stewards and officers. But as stated at the beginning of the chapter, each department is up to a point autonomous, run and controlled by its own senior steward be it of the mounted games, driving, quadrilles, or personalities. It would not be true to suggest, however, that Colonel Ansell as show director, left senior stewards to their own devices. While not interfering, he never hesitated to suggest improvements, criticise inefficiency, put his finger on any weakness. Indeed, as far as many departments are concerned the original conception of its functions and responsibilities were his.

For the most part, however, it is the senior stewards who plan the detailed running of the departments, and tremendously hard work it is. Their dedicated, conscientious, contribution to the Show is a principal reason for its success. They and their assistants give generously of their time, perhaps foregoing a week of their holiday, or more accurately two weeks, for it is the same team which runs the Royal International. Occasionally they are responsible for yet another show if, as sometimes, the Courvoisier Championships follow the Horse of the Year.

A few nowadays receive free accommodation; the remainder receive a small subsistence allowance, but few ever ask to be relieved of their jobs and few are dropped. By 1976, between them, the ten senior stewards had contributed some 230 years of service, or 3000 hours. Yet they are essentially the unsung heroes. To the general public, names such as Pride, Hurrell, Hall, Hudson, Brunt, Barnes, Bourne, all senior stewards with upwards of twenty years' experience, are completely unknown. Yet they are indispensable spokes in the Horse Show's wheel.

# 7 'They also Serve'

The hostesses—VIPs—sponsors and sponsorship—
the restaurant—timekeepers—accident drill—a
clanger from the band—Horse of the Year Show
directors of music—a probable protest

The previous chapter gives some idea of the comprehensive range of the behind-the-scenes responsibilities of the Horse of the Year Show, but the list is by no means exhaustive. There are many other departments, the activities and functions of which may not be apparent to the spectator but which are vitally important.

Each year, for instance, a number of riders from overseas are invited to compete. Some, of course, are old friends such as the d'Inzeos or the Schockemohles, but many are paying their first visit to Britain. There is the likelihood that their knowledge of the English language may be sparse but it is essential that there should be adequate communication. Riders not only want to know exactly when and where a competition is taking place, they also want to know of meetings, of parties and of possible transport to London or the airport. The Show therefore employs two hostesses who also act as interpreters. It is their responsibility to see that the riders from overseas are fully acquainted with the time-table, rules, declarations, requirements and hospitality.

Not necessarily intentionally, foreign riders can be very elusive, making the hostesses' job arduous and sometimes frustrating. In addition to being linguists the girls must be patient, sympathetic and tactful. Their responsibilities also include the organisation of the international club, where all the riders can meet after the Show for a drink and a sandwich, and to get to know each other. The hospitality room plays a major part at the Show and is responsible for a few very late nights. As much as anything it helps create the atmosphere so much associated with the Show. Without it overseas riders could feel left out. Because of it they feel very much a part of the whole happy family. Indeed, one of the pleasant aspects of show jumping is the camaraderie that exists between riders of all nations. They not only enjoy each other's company, they also respect each other, frequently buy horses from each other, learn from each other. Each wants to win; each is determined to win every time he enters the ring; but though he may be disappointed he is never resentful of another's victory.

The usual arrangement when overseas riders are invited is for the Show to accept full financial responsibility for both horses and riders from the moment they land in Britain, but frequently, additional financial assistance is offered to alleviate the cost of crossing the channel—an expense that riders do not experience when travelling from one country to another on the Continent. Once in Britain every effort is made to see that guests of the Show have the happiest time and the fewest worries or problems possible. If there is something seriously wrong they always have access to the Show director, but as a rule it is the charming and efficient hostesses to whom they turn for help or advice.

VIP's invited to attend in the royal box must also be looked after. Someone must meet them on their arrival and entertain them in the VIP room before taking them across to their seats. When chairman of the Show as well as the director, Sir Michael Ansell obviously could not attend to this in person, but it all has to be organised.

No less important are the sponsors. During the last decade or so Mr Bob Dean, treasurer of the BHS, has made himself responsible for finding firms and organisations prepared to back a competition. It is not difficult, of course, to find a sponsor for a jumping competition which is going to be seen by 10 million viewers. It is not so easy to obtain sponsorship for an event in the afternoon which has only the slenderest chance of being televised. Nevertheless, Mr Dean has achieved what many a few years ago would have considered quite impossible in producing some twenty sponsors for the Horse of the Year Show. Collectively they have been responsible for putting up over £20,000 which means that the Show can offer extremely generous prize money. Today, when jumping no less than any other sport has become extremely competitive, leading riders and owners will only jump at shows where the prize money is good.

In 1949 the first prize for the final event of the Show was £50, and the puissance earned £80. Today it is £1250 and £500 respectively. Only the leading show jumper of the year in 1949 commanded a reasonable first prize of £200, considered at the time to be quite exceptional.

It is relevant to put on record here that Mr Dean and the British Equestrian Promotions Company which he heads is responsible for producing another twenty sponsors for the Royal International Horse Show worth some £30,000. In addition, £40,000 is put up by John Pinches for dressage, while the Midland Bank's commitment to combined training, one-day and three-day events, must exceed £100,000. Over £300,000 was contributed to show jumping alone through sponsorship. Naturally sponsors must be looked after. If they are allowed to

Not only show jumpers from overseas are entertained at the Horse of the Year Show. Nuno d'Oliviera from Portugal proved immensely popular in his 'equestrian ballet' (*Stonex*)

feel that the Show is only interested in their money then it is not surprising if they withdraw their sponsorship. Accordingly, there has to be organisation for their proper hospitality. The sponsor of the main event of the performance is properly invited with his party to the royal box. Obviously the number of people that can be accommodated here is limited, though as a rule two or three rows immediately behind the box are kept as an overflow.

Sponsors of a lesser prize may be invited to have a table in the restaurant overlooking the arena. With increasing numbers of sponsors, working out the allocation of tables in the restaurant is like a jigsaw puzzle, but it is all sorted out with great tact and to everyone's satisfaction, by Bob Dean, his wife Lilian and his personal assistant, Miss Elsa Parry. The members' amenities, too, have in recent years improved enormously thanks to the steward in charge, Ray Stovold.

There are many other individuals or departments who still have to be briefed and relied upon to ensure the completely smooth running of the Show. Every rider has to be weighed with his saddle as he leaves the ring to ensure that he has carried not less than the prescribed 11 stone 11 lb (165lb)—not an inspiring job for the clerk of the scales placed, as he is, just outside the arena, at the edge of the collecting ring. Yet it is surely an example of the wonderful spirit evoked by the Show that the present clerk is none other than the man who was a driver in the musical drive of the heavy horses for over twenty years.

Vital too are the timekeepers, though these are not members either of the Wembley or the Horse Show staff. A well known firm is engaged to provide the equipment which measures to one tenth of a second. Obviously it has to be regularly tested for a fault can lead to much trouble, as on one occasion was learnt at the Royal International Horse Show.

It was at the Royal International, in 1975, that for the first and only time a crucial piece of drill was put to test and was found not to be wanting. There is always the possibility of disaster befalling horse or rider. As far as the latter is concerned there has been no shortage of incidents over the years, though none, fortunately, fatal. On the first morning of the Show, or during the previous day, a drill is worked out with the medical officer, St John's Ambulance orderlies and stewards, and is meticulously rehearsed. On such occasions when it has been necessary to put it into practice the drill has worked without a hitch. Fortunately these have been few and far between—riders being the tough and plucky people that they are generally manage to get to their feet after a tumble.

However, nature being what it is—at least as far as the British are concerned—the crowd and, I believe, people watching on television—are sometimes more concerned if a horse is damaged than if a rider is hurt. This did not appear to be the case in 1953 the year after the British team had won an Olympic gold medal at Helsinki. At the very height of their popularity the great Harry Llewellyn and his immortal Foxhunter took a crashing fall at a double at Harringay. Hitting the second part Foxhunter caught one of the poles between his legs, stumbled, pitching his rider over his head and then appeared to roll on him. Foxhunter was up in seconds, but Colonel Llewellyn lay still. After a moment he half raised himself to his knees, then dropped forward and appeared to be fumbling his way towards the side of the arena. Two orderlies of the St John Ambulance team rushed forward, but Harry waved them away and continued to grope his way to the side. By chance this was right in front of the commentary box. Looking up he called out to me: 'I'm not hurt, I'm just looking for a tooth I've lost: you'd better tell them.' After the moment of anxiety the roar of laughter at the explanation nearly brought the roof down.

The distress for a damaged horse was exemplified when a horse was tragically killed in 1975. It also emphasised the need to have a properly rehearsed drill. In fact, enormous care is taken over this. The moment that there has been an accident the lights are dimmed on an order from control; control, too, calls for music, from the band if it is in position, or from the sound room where a record or cassette is always ready. At least one veterinary surgeon is permanently in attendance, sitting by the ringside, as is a doctor. Ambulance orderlies are situated at each corner. In one corner there is an 8ft high canvas screen. The arena party knows exactly where it is, and it is their leader's responsibility to call for it the moment that he is aware that it is needed. Outside in the collecting ring there is a low trailer with a winch and pulley, already attached to a Land-Rover with a driver standing by.

The drill is rehearsed thoroughly at the beginning of every Show; and before each performance starts there is a check to make sure everything is in its proper place. At the 1975 Royal International when a horse was killed in the King George V Cup, Alan Ball, the course builder, saw at once what had happened and his signal to the collecting ring for the trailer alerted control, the band, the veterinary surgeon, the doctors and the St John's Ambulance Brigade. Everything worked like clockwork. In just 90 seconds the next horse was in the arena and ready to jump, the knocked-down fences having been rebuilt, the trailer with its sad load pulled out, screened by the canvas held in place by members of the arena party.

The show must go on: lights up, the band is stopped, the bell is rung. The whole operation was so quick that I had to ask the

Henri Gilhuys from Holland with his haute école was warmly received by the Show's audience (*Stonex*)

judges to hold up the start of the next horse as the short film that the BBC had switched to had not finished. There is no compensation for a tragedy of this nature. That the necessary drill can be carried out so expeditiously is testimony to the efficiency of proper rehearsal, even though there is hardly ever a need to put the drill into operation. Yet one day there may be a disaster, and everyone must be ready for it.

But even the best laid plans can go awry. The procedure for the arrival of royalty at either of the big shows is always the same. When the royal personage arrives at the stadium, control is informed. Some five minutes later after presentations in the VIP room and perhaps a little refreshment, a move is made to the royal box and again control is warned. If a jumping event is in progress then control tells the collecting ring to stop the next horse. If it is a show class then the band is told to stop playing, which is the cue to the ring stewards to stop the class.

As the member of the royal family reaches the entrance to the box from the alleyway at the back, a physical signal is given to control. On reaching the centre of the box there is the announcement: 'We now welcome to the Horse of the Year Show for 197— Her Royal Highness . . . .' The band immediately plays the salute, the full anthem or the first sixteen bars according to rank.

At both Wembley shows this procedure has gone without a hitch for years. On one occasion there was a lapse. The Royal personage did not arrive at Wembley until some half hour or so after the start of the performance. All the usual drill was brought into operation. Her Royal Highness reached the box: 'And now, ladies and gentlemen, we welcome to The Horse of the Year Show—.'

Everybody stood with the usual clattering of seats, then— silence! No-one had noticed that the band had left its stand, as it usually does once a jumping competition has started, no music then being necessary. Usually the band sergeant rings control to ask approximately when they will be wanted back and are asked to be in position about 10 to 15 minutes before the end of the event. On this occasion both the director of music and the band sergeant had forgotten that they had to remain in position until after the royal arrival. After what seemed a very long moment Her Royal Highness, having glanced across at the empty band stand, sat down: as did everyone else.

Sir Michael was not amused. His whole life is based on military precision. He demanded the immediate presence of the band master who very soon reported at control, expecting no doubt the biggest rocket of all time.

'Well?'

'I'm sorry, sir. I . . . er . . . I just forgot about the salute.'

'Well, I bet you'll never forget again.'

'No, sir.'

'Right. That's all.' . . . except for the leg-pulling at the conference next morning. As Sir Michael so rightly realised, the shock of the enormity of his incompetence would be far more salutary than any harangue. But I have no doubt that the band master was relieved at being let off so lightly.

It is because a lapse such as this is so rare that it is so memorable. In fact, the band at the Horse of the Year Show plays an extremely important part. It not only entertains, it accompanies the Show. High standards of musicianship are demanded of it and also a high degree of intelligence and a quick wit from the director of music.

After the disastrous organ of 1949, the first band to be employed was the Morris Motors Band. It was not surprising, however, to learn the following year that Mike Ansell had settled for a military band. He chose an accomplished and extremely cooperative band provided by the Royal Army Service Corps. Their director of music was a charming and amusing man, Captain J. F. Dean, who threw himself enthusiastically into something that was then quite novel: not only the Show itself but the concept of the contribution that could be made by the band. He quickly endeared himself to the whole team and in 1962 earned the good wishes of everyone when he was promoted to the rank of Major.

In our ignorance in the early days we frequently referred to him as bandmaster instead of director of music, but he always took it in good part. He was full of ideas and three or four months before the Show he would ring me up from Aldershot to tell me of any ideas that he had for the tunes for the personalities, and wanted to know of any suggestions that I had. He would also discuss the music for the displays and would happily, and quite unselfconsciously, sing down the telephone tunes the titles of which meant nothing to me. I would sing or whistle tunes back to him when I did not know their correct name. The telephone exchange must have found it all rather bewildering, but we usually came out with some good ideas which delighted or amused the team, even if, on occasions, they were quite lost on the audience.

News of the sudden death of Major Dean after the 1962 Show came as a great shock. He had become so very much one of us.

In 1963 his place was taken by a Lieutenant Desmond Walker, who was tall, smart and impressive. He had risen from Kneller Hall to be director of music of a first-class regimental band and was to go further. After only two or three years he was promoted Captain and in 1970 he became director of music of the Welsh Guards, his band immediately being engaged to play at the Royal

*above* No visitors are more popular than the d'Inzeo brothers: Piero is seen here on Pioneer (*Roberts*)

*right* Hugo Simon, from Austria, here seen on Flipper, is another visitor to Wembley who always gives value for money (*Lane*)

International Horse Show. Shortly after he had heard that he was to be senior director of music for the whole Brigade of Guards, which would lead almost certainly to the command at Kneller Hall, he had a fatal heart attack in his hotel room during the 1974 Royal International. It was a great loss, but none who attended his Memorial Service at the Guards Chapel will forget how that superb musical occasion reflected his talent and his personality.

By this time Captain 'Bill' Allen had taken over at the Horse of the Year Show, his band now being the Royal Corps of Transport. He learned the ropes very quickly and adjusted himself to the unusual role expected from the band at Wembley. He never puts a foot wrong, and like Dean, he is both co-operative and inventive.

The drill of the band is now traditional: the lights are slightly dimmed as it starts playing 15 minutes before the start of the performance. At two minutes to the start they crash into the Regimental March of the Junior Leaders Regiment, as the arena party smartly march in. When they are halted in the middle the band changes to the Show's signature tune *Greensleeves*. As the

tune nears its end the start of the Show is announced.

The band then stand down, unless royalty is arriving, until the end of the competition. Immediately the rosettes have been presented they play for the lap of honour always trying to link the music to the winner: *Ilkley Moor bar tat* for Harvey Smith, a snatch of some national music for the winner's nation; occasionally a joke—*Humpty Dumpty* for Ted Edgar. Experience enables them to produce almost anything at a moment's notice. It may be that there is a hold-up for some reason—an injury, a completely broken fence, a light failure—then immediately the band is expected to spring into action.

For the big displays, the dressage, the musical drive, the personalities, all the music is carefully arranged and rehearsed weeks ahead. It can add enormously to the effectiveness of a display. The band's own moment comes, of course, in the interval when, as the lights go out on the Trumpeters and the Household Cavalry at the end of the parade of personalities, the band is hit by the spots and the audience is informed of the music they are going to hear. It is usually a selection from one of the great musicals, but whereas at the Royal International it might be *South Pacific*, *The King and I*, or even *The Desert Song*, at the Horse of the Year Show it is more likely to be *Jesus Christ, Superstar* or a selection of Bert Bacharach's tunes. The Horse of the Year Show attempts to be a little more up-to-date (though the Christmas Show at Olympia tries to be more with-it still).

Often, if the Show is running a few minutes late the length of the interval is reduced. With a herculean effort the course builders get the new course up in eight minutes; the riders are called in to walk the course; the bell is rung by the judges to let them know that they have had long enough and must clear the arena. Control enquires if the first horse is ready. It is. 'Stop the Band!' Unceremoniously the band is silenced almost in the middle of a bar. But no one complains. This is the Horse of the Year Show: it is run to split second timing and everyone is expected to accept this fact. Invariably they do.

For an hour the band stands down until the end of the major competition: or, perhaps, just the end of the first round, music being required while the course is altered for the second round or jump-off. Then it is the music for the lap of honour, and the *National Anthem*.

Even that may not be the end. There is a different dressage display the next day and the performer has been given permission to use the ring for a practise half-an-hour after the end of the Show. The band is required. Uncomplainingly and conscientiously, it accompanies the rehearsal until it might well be midnight. Then the bandsmen get in their coach to drive back to

Aldershot. Yet to the constant benefit of the Show they never flag or grumble. At 2.00 pm next day they play the march in with as much zest and attack as if they had never played it before.

Did someone recently demand that military bands should be done away with? Such a move would win little approval at the Horse of the Year Show. A full scale protest would be more likely, which, led by the show-jumping boys, could be quite formidable.

# 8 The Source of Inspiration

Morning check-up—the escort—instructions for the day—the morning conference—post-mortem—details of the day—a working lunch—after the Show is over—late-night listening

The disparate parts of the Show organisation are all welded into one by the morning conference which is held every day of the Show at precisely 12.00 midday, invariably chaired, until his retirement, by Mike Ansell. But long before this meeting the Show had been operative, and Mike or one of his chief assistants has checked on anything requiring attention in any particular department, be it complaints about the surface in the outside arena, some trouble in the caravan park, or an objection made against a competitor.

Until recently Mike would be on the show ground by 8 o'clock in the morning, be led round on a tour of inspection and then have breakfast at the arena. Being blind he naturally had to have an escort and over the years he has always chosen someone with whom he can discuss matters and whom he can trust implicitly. In the early years it was General Vivian Street, then Charles Stratton; more recently Colonel Guy Wathen, for many years an international jumper, and also at one time in Colonel Ansell's old regiment: finally it was Paul Mercer.

This escort has had an important role to play, acting as Mike's eyes, keeping him informed of any number of details that must inevitably escape a blind man: 'The flowers in front of the Royal Box are looking a little faded!' 'The sign over the entrance was crooked for the display last night.' 'A number of cars are parked in the drive to the stables.' In addition, the escort has had to be able to protect the director from extraneous details or unnecessary trouble: people trying to pester him with details irrelevant at that time; a petty grumble by a competitor or steward, a row between two exhibitors.

But this can put the escort in a quandary. Colonel Ansell has always wanted to know everything that is going on. With the best intentions in the world the escort might conceal something which Mike might learn about later. There could then well be an explosion from Mike:

'Why wasn't I told?'

'I just thought . . .'

Rebuilding the famous Puissance wall. Alan Ball, the course builder, can be seen peeping over the top (*Davidson*)

'You're not here to think. I should have been told at the time!'

But these occasional outbursts were seldom resented. There might have been a certain childish schoolboy camaraderie among the stewards, supporting colleagues in trouble with the headmaster, but the authority of the headmaster was totally accepted—and respected. All were completely aware that his one aim was to make the Show the most successful of its kind in the world. It was equally the one aim of the team to assist in this, and if from time to time someone was reprimanded, whether wholly justified or not, no one—or very few—ever resented it. For one thing it was appreciated by one and all, that no one worked harder than 'the boss' himself.

Latterly, understandably, he did not appear at the Stadium until about 9.30 am, and rested for a couple of hours in the afternoon; but after so long the senior stewards were more than able to see that their own department was running smoothly, even at 8 o'clock in the morning, or while a matinée was actually in progress. As far as the overall responsibility for administration is concerned the mantle of Mike Ansell has now fallen on John Stevens.

When running the show, as soon as he arrived at the Stadium Mike would go to his office where his secretary would be waiting with the mail. For two hours he would be closeted there, a major part of the time being taken up with the working out and dictating the instructions of the day. These would be meticulously worked out—as they still are—and would have to be completed in time to be typed and run off in the general office and distributed among those attending the conference.

At five or ten minutes to twelve stewards start assembling in the conference room. Within a minute or two of 12.00 pm Mike Ansell would be piloted into the room, making his way along the back of the chairs all down one side of the table. With all the senior stewards or their representatives and members of the Wembley staff the daily conference comprises some thirty people: there are never quite enough chairs. Having reached his place at the end of the table, flanked on one side by his escort and John Stevens, and on the other by myself, he would take his seat, having put his stick under the table, ask someone to get him a drink, and, feeling his St Dunstan's watch, enquire the exact time.

'Right: we'll make a start.' He hardly had to raise his voice: everyone was ready and waiting. The morning conference, from which generates all that makes the Show tick, would be under way.

'Before we go through the instructions I'd just like to thank you all for making last night such a superb night.' Or, possibly:

'We ran late last night, for which I apologise, but a number of things went wrong.' And he would detail them. The comments could be sharp or sympathetic, but no one was in any doubt that he was fully cognisant of any weakness in any department.

'It's not necessary for all four hacks in the championship to give a show when we're short of time. ... Why couldn't you introduce a standard? ... You should have stopped the judge going right down the back row ... Why didn't the band play a snatch before the first jump-off?'

Excuses or explanations are forthcoming, but they are not laboured. There is no point.

'Any problems with the dressage display rehearsal this morning John? Right. How many St John's people were present when they started jumping at 8.30 pm? Only two: there should have been four. Chase them up, will you David? Did they keep up to time with the ponies? Well done, Barry. Who's here from Wembley? Will you thank everyone for having the place so clean? Are the foreigners happy? I'm told the loudspeakers were hard to hear in the members' seats. Can you check? Did anyone else find the band too loud? In the restaurant I could hardly hear myself think. Have you sorted out that problem with Mrs So-and-so's stabling? Good. Who's in the royal box tonight?'

And so on: any loose ends are tied up; minor problems are sorted out; any weaknesses are rectified.

'Right: now for today. Read it out, John, will you?'
In detail, the day's instructions are read out.
'Two o'clock, arena party march in. Gordon Richard Stakes. Estimated thirty to jump: one jump-off against the clock.'
'Wait a minute. How many have declared?'
'Twenty-eight, Colonel.'
'How long have I allowed?'
'Sixty-five minutes.'
'You'll be fairly pushed. Keep it going, and we don't want too many clears. Six clears at the most, Alan.'
'Yes, sir.' (Alan Ball, the course builder, who is not only expected to build a course that will produce the required number of clear rounds, but usually succeeds.)
'Right, carry on.'
'Small Hack of the Year. Eight to come in: first two to give a show: all get rosettes. Cup to be presented by Mrs . . . '.
'Do you know which table she's sitting at?'
'Yes, sir.'
'Well get her down in good time: the judges can do the rosettes for the rest. Will we have spotlights by then?'
'Yes, sir.'
'Then we can spot the winner, but cut them out if they start playing up.'
'Three-twenty, Mounted Games.'
'Which games have you got this afternoon? Well keep pushing. I want to make up ten minutes if I can as there are more declarations for the Young Riders than I expected. Jack will you give the rosettes?'
'Four o'clock: Personalities. The race-horse won't be coming in.'
'Then you ought to make up a few minutes there.'
'Four-twenty, Interval.'
'What are you playing?'
'*South Pacific.*'
'First class. How long have I allowed for the Interval?'
'Twenty minutes.'
'We'll try and get it through in fifteen minutes. I want to finish the afternoon by 5.25 pm.'
And so it goes on. Every detail, so that no one can possibly fail to be aware of the overall picture of the performance. Then it is the evening programme, in even greater detail, because it is of greater importance that there should be no hitches, that the timing should be kept accurately, if only because at 9.30 pm television starts.
'What are you doing this evening?' (to the television producer, always present at the conference.)

'We're recording the William Hanson Trophy and then we thought we'd do the quadrille. We'd like to join the Sunday Times Cup at ten past ten, in time for the first jump-off.'

'How many did I say we wanted in the first jump-off?'

'Eight and then four.'

'Twelve rounds in all. So you ought to get the presentation.'

'That's what we'd like.'

'Right. You'll have it.'

Of course, things don't always work out so exactly and the timing of the Show can go wrong, two or three minutes being lost over each item: occasionally more clear rounds are jumped than expected, sometimes it is the fault of the show director himself, too little time has been allowed for a display, or a competition; sometimes it is the course builder who has to take the blame; there may be twenty clear rounds instead of five; a judge may take too long; there may have to be a run-off because of a tie in a driving competition.

Late running is not popular with the Show—over-runs are anathema to the BBC. It is thanks to the excellent relations that have been built up between the BBC and the Show, and of course to the popularity of show jumping with the viewing public, that 'presentation' (controller of all transmissions) has frequently agreed to over-run rather than insist on cutting off a programme before the end. Every possible effort, of course, is made to ensure that the timing of the Show fits in with the TV requirements.

The conference over—'Get me another gin and french!'—everyone returns to their job before taking a hurried lunch at the Silver Mint. One or two are asked to stay behind to discuss some special problem that need not involve the whole assembly. Finally, Mike Ansell with his escort, John Stevens, myself and probably General Jack Reynolds, the director-general of the British Equestrian Federation, who is responsible for the organisation of the royal box, go up to the restaurant for a working lunch, during which Mike liked to think aloud, to discuss the general running of the Show, and to look ahead. Occasionally someone would be invited to join the table, but Mike would make it clear that he could not pay them much attention from a social point of view: even when feeling relaxed the conversation—and the jokes—would seldom be concerned with anything but the Show.

For many years this was the only proper meal that we would get during the whole day, apart from a few sandwiches available in the conference room in the interval, or at the end of the Show. In later years, however, Mike Ansell would go up to the restaurant after the interval in the evening performance, when the last class,

A great jump over the big wall by Caroline Bradley and New Yorker, a famous Puissance combination (*Davidson*)

the main competition of the session, had started, and have a light dinner. For me, as often as not, it was still sandwiches, for invariably the main competition of the evening was being televised, and so I would be on duty.

Half an hour or so before the end, Mike would resume his place behind control.

'How many clears? How many more to jump? Are they going to bring in a standard? What time did television start? Tell Alan I want to know the heights in the jump-off. The band will have to play a snatch. Raymond (or Chris), don't forget to thank the arena party: their Colonel's here tonight. Remember to mention the sponsor. Who's on the PA for the awards? Just listen to that! (applause after a clear round by a popular rider) Tell George it's better than the Cup Final.'

At last it's all over. 10.30, 10.45, 11 o'clock. The anthem. The exit march—*Fare thee well, Inniskilling*, if it's been a good night. We all retire to the conference room for a drink and a sandwich, a chat and a laugh.

'Who's in here?'

'Only the stewards and the judges.'

Not always strictly true. In the interests of economy too many friends and guests are frowned upon but it is difficult for Mike to check. Nevertheless the point is taken.

'I'm going to bed. Just look in on the VIP room, Dorian. Tomorrow ought to be the hell of a day.'

Yes, already we're looking toward tomorrow. Indeed, it is not far off. Yet probably most stewards, despite their tiredness or perhaps because of it, are late to bed rather than early: the end of term feeling is still dominant. After all one will not see a lot of these colleagues with whom one is working so closely for another six months or more.

'What are you drinking?'

It's the same in the caravan park and at the Foxhunter Bar.

'Well, only one more, and then I really must go to bed.'

'What time's the first class in the morning?'

'Eight-thirty.'

'Christ! Good night!'

In his room at the hotel Mike is listening to the midnight news—just to make sure that they've got the result of the Sunday Times Cup on it. Somebody will be in for a raspberry in the morning if they haven't, especially if it was a British win.

# 9 So Long in the Saddle

Colonel Sir Michael Picton Ansell—early years—his career as a soldier—wounded at St Valery—his flowers—call from the BSJA—the Victory Championship— manager of the International knighthood—a long innings—no thought of giving up—dedication to detail—twenty-first criticism 'You're not here to think'—will to *win*—the undisputed stature—a tribute

No one in any way associated with the Horse of the Year Show—until 1975—could ever be in any doubt as to who was in the saddle. Colonel Sir Michael Picton Ansell towered over the whole Show both in physique and in authority. Though he always denied it, his was in fact a benevolent dictatorship; when on occasions it was something less than benevolent, then as a rule there was justification for what could be described as autocratic behaviour.

Michael Ansell was born in 1905. His father, a regular cavalry officer, was killed, leading his regiment, the 5th Dragoon Guards, in September 1914. As noted in chapter 2, my father, Colonel V. D. S. Williams, his equitation officer, became his guardian. Mike never had any desire to be anything but a soldier. In 1923 he joined the Inniskillings which a year earlier had amalgamated with the 5th Dragoon Guards. He quickly made his mark in every sphere: he represented his regiment at polo, and later his country; he was a very successful point-to-point rider, he became his regiment's equitation officer, he represented the army in show jumping, winning his fair share of prizes in international competition; and he produced the famous trick ride which achieved great fame between the wars. Especially popular was the jumping section when, led by Mike Ansell on his legendary horse Leopard, they jumped just about everything that could be jumped, without stirrups or reins, including a line of swords.

In the late 'twenties he was selected for a course at the Cavalry School at Weedon. As Weedon was only a few miles from our home near Towcester, for that year he almost lived in our house. It was a memorable part of my childhood. When Mike was around life was always exciting: or, as he would say, *fun*. He revelled in leg-pulls, teasing, practical jokes and hoaxes. There

was an almost non-stop banter between him and my father, and frequent horseplay. Looking back, those ten months seem to have been months of non-stop laughter.

At that time my father was Master of the Grafton, one of the packs with which the officers at Weedon hunted. There were few better men across country than Mike Ansell that season, even though it was a vintage era. As my father used to say, for the Master to keep in front in those days was no mean task, for there were some of the finest horsemen in the country at Weedon, led by the instructors which included Arthur Brooke, Jock Campbell, 'Bede' Cameron, 'Friz' Fowler: all well known in the annals of the cavalry.

By the mid-thirties Ansell had made a very considerable reputation for himself. In 1940 he was given command of the Lothian and Border Yeomanry—at 34 the youngest commanding officer in the army. Being ambitious he was naturally delighted, though sad to leave the Inniskillings which he had always hoped to command. His contemporaries at the time speak of him with admiration, respect and, of course, affection; though many were aware of the ruthless streak in him, and realised even then that if Mike really wanted something, practically nothing would stop him. His determination, always, was to *win*.

There is a story told of how when the Colonel came round to inspect the recruits' ride he was greatly impressed by those in Captain Ansell's squadron. His brother officers were not altogether amused when they learnt that he had substituted experienced riders for the recruits! In his book, *Soldier On*, he tells how when he learned that under new regulations shortly to come into being no officer under 35 could be promoted to the rank of major, even to fill a vacancy, he persuaded a fellow officer who was leaving to send in his papers a month early, for which he paid him a month's major's pay. He thus got early promotion: as he says in the book, 'a very good deal'. Certainly, but it showed him to be an opportunist, always ready to seize a chance, having had the foresight to be ready for it.

In 1936 he had married Victoria Fuller and with the birth of his elder son, Nicholas, in 1937 and his unusually successful career with its exceptional promise, the world appeared to be at his feet. But in 1939 war broke out. At St Valery a few months later he received the wounds that were so to alter his whole life.

For many months he was 'missing'. Then when in hospital in Paris he persuaded a visiting Irish priest to get in touch with Dan Corry, a doyen of the Irish show jumpers and now frequently a judge at the Horse of the Year Show, and ask him to let Victoria know that he was alive. Eventually he was repatriated after a long spell in a German prisoner of war camp, time not entirely wasted

Colonel Mike (*Davidson*)

as he spent much of it planning the revival of show jumping in Britain. He still retained a little sight in the corner of one eye, 'guiding vision'. It was made clear to him that if he lived a quiet and sedentary life his limited sight need not for many years deteriorate: but if he lived a 'normal' life it would go.

Inevitably, of course it did go. It happened at the time when the *Daily Mirror* was putting on a Cavalcade of Sport at the White City. There had been a rehearsal the evening before, after which I had had dinner with Mike at the Cavalry Club. When I collected him next morning he said, 'I think I've had it. I couldn't find my razor this morning.'

Months attending St Dunstan's had prepared him for this, so that for most people Mike's transition from limited sight to blindness was virtually undiscernable. Even today many who have been intimately associated with him for years find it easy to forget that he is blind. They hold out a hand, expecting him to take it; they pass the salt to him across the table, expecting him to know where it is; they assume that he knows who it is that has entered the room. Frequently, in fact, he does, because he recognises a voice instantly, even though he may not know the person very well. Often—deliberately, it seems—he uses the words 'see' and 'look': 'I see you're wearing a new dress.' 'Don't the flowers look marvellous?' Like so many blind people he has a remarkable memory and has only to be shown the geography of a room, or indeed a house, once for him to be able to find his way around: though there is no doubt that in recent years his reluctance to hold the Royal International anywhere other than at Wembley was because he knew Wembley so well: he was not prepared to 'learn' a new environment.

Mike frankly admits that on occasions he has turned blindness to an advantage, to get something he wants, to achieve something through sympathy. On one occasion, he even told me that if, miraculously, he learned that his sight could be restored he would reject it. His whole life, he said, was now orientated towards being blind: sight, all that it entailed in so big a change to his way of living, would only be confusing and disillusioning after so long. Once again, in his extraordinary manner and thanks to his remarkable fortitude, he has managed to make a virtue out of a necessity. A few years ago after I had survived a serious illness he wrote to me, 'Remember fortitude is so much greater than courage.' This, of course, is what he has himself proved.

With his customary resilience, once home again he sought out some kind of creative activity. Flowers had always been his second love and so he decided, with his wife Victoria, to start a flower market-garden—which has, incidentally, been a very considerable success—at Bideford in Devon. This, probably,

Princess Anne assisted by Sir Michael Ansell cuts the twenty-first Birthday cake (*Lane*)

would have been his full-time job had he not been persuaded to stand for the chairmanship of the BSJA. In December 1944 he was elected by one vote. What a crucial vote that one turned out to be.

In September 1945 he organised, at the White City, the very successful Victory Championship at which he was able to put much of his prisoner of war camp thinking into practice. In 1947 when the International Horse Show was revised, also at the White City, my father as chairman asked him to be responsible for the organisation of all the jumping events. In 1950 he became show manager, having a year earlier launched the Horse of the Year Show.

For twenty-five years, therefore, he has dominated the two major or 'official' shows. In addition to this he was for over twenty years chairman of the BSJA and for nearly as long chairman of the BHS, having previously been honorary director. In 1972 he became chairman/president of the British Equestrian Federation, retiring in April 1976.

For anyone, let alone a blind man, this has surely been an unusually full and rewarding life. Yet is a man of Mike Ansell's character ever wholly rewarded, entirely satisfied? It is a curious trait in the personality of men of ambition, even the greatest, to feel that they have not achieved quite all that they should have. There are endless examples in history of men who have attained the highest office, yet have died disappointed. Recently in the diaries of Lord Reith, father of the BBC, it is evident that although he achieved an eminence that most would feel as great as any one man could expect, he felt frustrated not to have reached even greater heights.

When in the New Year's Honours of 1968, Mike received a knighthood, many believed that he would then retire, or at least allow himself gradually to be relieved of some of his work. He was nearly 65; he had been running the two shows for 20 years; he had held for a considerable time the highest offices in the horse world; in addition he was, at that time, suffering considerably from back trouble. He had certainly earned an easier life. But, in fact he appeared to attack the 1968 Show with more than usual vigour, as if to make it quite clear that he had no intention of giving up, or perhaps to prove to himself that he was not going to be beaten either by his bad back or increasing tiredness: once again he had to *win*.

It was noticeable, too, that he paid even more attention to detail, delegated even less. One example from many: a presentation was to be made by the Pony Club to a children's charity. General Vivian Street was to receive the presentation at the foot of the royal box. Some twenty children were to ride in, each carrying a bag of money. At the morning conference I was detailed to organise this and a rehearsal was to be held after the afternoon performance at 5.30 pm. As the performance ended early and all the children were ready by about 5.20 pm I went ahead with the rehearsal. The ring was to be in darkness; suddenly the twenty young riders all in their coloured shirts, were to explode into the arena as the lights blazed on, gallop round and draw up in a posse before the royal box, throwing their bags onto the floor.

We had just finished the rehearsal when Mike Ansell was led into the ring. He came across to where the children were congregated and started to address them, telling them what to do and so on. After a few moments I interrupted, telling him that I had arranged it all, that it was all over. He was obviously not pleased. Why had we started before 5.30 pm? I explained. What exactly had been organised? I told him. Had I not added that Colonel Guy Cubitt, chairman of the Pony Club, was there and had approved it all I believe he would have insisted on taking charge himself and re-rehearsing. He would never abdicate complete responsibility.

He would argue—perhaps with some justification—that a great leader leaves nothing to chance: yet it has also been said that too great attention to detail can be a weakness in a leader. In the case of Mike Ansell it was, quite simply, that he saw his attention to detail as part of his image. It is his philosophy that it is to the benefit of any group or body or organisation, if there is only one personality at its head, only one person surrounded by an aura.

If a structure is to be formidable and effective the driving force must emanate from just one person. Was this not the case with Churchill and Monty? Mike Ansell did not consciously suppress those efforts of others which might gain them kudos, increase their stature: rather, as Lord Hailsham once said of Edward Heath, he was like a massive oak that by nature allows little to grow underneath it. That this was an entirely subconscious attitude there can be no doubt. Privately he was often lavish with his praise: there are few senior stewards who have not experienced evidence of his gratitude.

He was hurt and surprised, and therefore angry, when I suggested after the twenty-first Horse of the Year Show, that it would have been a gesture much appreciated if he had in some way singled out and rewarded others who had been associated with the Show for the whole twenty-one years, as he had: invite their wives into the royal box one evening, give them a table in the restaurant, or ask them to present a trophy. That he had not thought along these lines was not deliberate: it just had not occurred to him that such a gesture was necessary. He was, deep down, immensely grateful for the help that he had received from these senior lieutenants, but he knew that their contribution was only contingent upon his.

On one occasion a visitor was heard by a senior steward to remark to Mike that he must have a wonderful team of helpers. His reply, instinctively, was that it was a wonderful team 'because they bloody well do as I tell them'. The senior steward was annoyed until he appreciated that this was Mike, the essential character of the man. He genuinely saw his stewards' efforts as reflections of his overall design: and he was right. Every department reflected his planning, his thinking, his inspiration.

At the dinner to celebrate his knighthood, in proposing his health I reminded the distinguished gathering that we had all enjoyed the experience of coming to him with some idea which he had immediately rejected, only to produce a few weeks later as his own! I was, of course, pulling his leg and no one laughed louder than Mike, except perhaps Prince Philip. But the germ of truth in my joke had been appreciated by all present who recognised his occasional tendency to take credit to himself, rather than allow it to others, but a characteristic that few resented.

Surprisingly, he has, on occasions, shown himself to be extremely sensitive to criticism, especially from the press. More than once, when a correspondent has been critical in his or her paper, instead of ignoring the matter as his closest associates have advised, he has retaliated: first by referring to the matter in scathing terms at the morning conference, then by summoning the offender and giving him or her a sharp, sometimes bitter rebuke. It has to be admitted, however, that once he has relieved himself of his anger he has behaved towards the offender with charm and generosity. Once he attempted, unsuccessfully, to persuade an editor to dismiss his equestrian correspondent. On another occasion he went out of his way to attack a correspondent in front of other people at a reception, because he had taken offence at a headline, which, of course, was not the correspondent's responsibility. He would have been surprised if he had known that he had offended the correspondent in question. He had not meant to hurt, just to get his own back, to achieve the last word.

'You're not here to think', became a joke phrase among stewards for several years after a senior steward had been reprimanded for giving what was, in Mike's opinion, wrong advice to one of the displays:

'I'm sorry, sir, I only thought. . . .'

'You're not here to think.'

To his credit the steward retaliated:

'No, sir, and I'm not just here to be a yes man.'

Which, of course, Mike Ansell appreciated, and respected.

Why, then, these flaws in a character that has so easily and convincingly won universal admiration and respect? Let it first be said that the great man who has no flaws does not exist and never has. The portrait of a man, at least in words, that is anything less than 'warts and all' is false. The very greatest are human, and humans are subject to the frailties of nature.

Having said that, it does seem that in a surprising number of great men there is a hidden, almost inexplicable lack of confidence in themselves. Why this should be so is difficult to understand. It is as though, despite the apparent success of their lives, they feel that they have not achieved all that they might have done. Something has eluded them: so they must press on, continue to prove themselves to themselves, and to the world, whatever the cost: there is ample evidence of this, even in Napoleon.

At the end of his autobiography, Mike Ansell writes: 'Many would say that I've had more than a glimpse of fame and glory: but the iron fulcrum has turned my fate another way and complete happiness eludes me.' Fate has certainly dealt him cruel blows, but perhaps there is more to it than that—it is not just a matter of his personal happiness. Subconsciously, he must be aware that were it not for his disability he might have become chief of the general staff, the highest position in the Army. He must feel, too, that with his proven drive, ability, clear thinking and personality, he could have been a captain of industry, in the £40,000 a year bracket; he could have been at the top of almost any profession that he chose. Somewhere in his make-up, therefore, it is almost inevitable that there is a sense of frustration. In settling for the

Line-up for the Police Horse of the Year, an unusual but very popular class. Sir Michael Ansell can be seen centre having congratulated the winner, PC Dracup from the West Yorkshire Constabulary (*Lane*)

horse world—or having had it thrust upon him in 1944—he accepted a great challenge. It must be accepted as of no less importance than any other of the worlds that he might have conquered but for his disability. It was a challenge that he had to *win*: 'We've damn well got to win.'—how often one has heard him say that.

And he *has* won: but to win this particular battle it has been essential for him to dominate. On occasions his domination may, unintentionally, have hurt or offended: but it has seldom been resented because everyone in the horse world has appreciated that at its head was a man with the stature to have been head of almost any organisation in the country. We have been proud of this, particularly, perhaps, those of us concerned with the Horse of the Year Show for it is this Show more than anything else that has popularised the horse with the general public.

In 1954 Mike Ansell asked Ronald Duncan to write that famous tribute to the horse, *Where in this wide world?*, which has perhaps become more identified with the Horse of the Year Show than anything else. In 1975 I suggested to Ronald Duncan that he might write a tribute to Mike Ansell for inclusion in this book. To my delight he agreed.

>Taller than his shadow: a man
>>who is patient with servants,
>>impatient only with his friends;
>>we like him for his virtues,
>>love him for his faults. A man
>>who knows the difference between fortitude
>>and courage; discipline
>>and obedience. Who
>>to a rude age brings a consistent gentleness.
>>His perceptive hands sign kindness
>>on flank, girth or flower.
>>He, grateful for our sight; we for his vision.
>>There is a bright candle burning in his mind.
>They say he is blind.

Blind, yes: but he 'saw' the Horse of the Year Show.

# 10 Fun and Games

The Cat and Custard Pot—Sergeant Major Lee to the rescue—the Surtees centenary—the Vale of Aylesbury Steeplechase—punctured pride—St Cyr, von Nagel, Schultheis—Lise Hartel—Peralta produces a few problems—Chamartin and Fischer—Oliviera, Gilhuys and more criticism—the hackney carriages—the mounted games

Nothing has been more popular with the general public at the Horse of the Year over the years than the displays. Particularly in the early days of the Show, there were some unforgettable items. The first really ambitious display was in 1954. To be called The Meet at the Cat and Custard Pot it was to be presented by the Pony Club. The Cattistock and other West Country Pony Clubs were largely responsible as the Bullen family, then living at Catherstone, produced most of the ponies and many of the riders. There was also a harness contingent from the Wimbledon Pony Club and a party on foot from the Old Berks. Altogether eight Pony Clubs were represented and the whole display comprised some forty young people all under 16, except Mrs 'Pug' Whitehead who, although an international rider herself, was small enough to ride a pony and act as the famous huntsman, James Pigg, the North Herts Beagles providing the hounds. With all the children on ponies and no grown-ups taking part it was a complete hunt in miniature.

It may have been the first of the Horse of the Year Show's big displays. I am inclined to think that it was the best, certainly the most charming. I had been asked to prepare the script and produce the display which turned out to be a rewarding experience. As the lights were lowered and the inn sign was brought in the band played very softly *Just a Song at Twilight* and I started speaking: 'Autumn, and with the long winter evenings we pull up our chair to the fire, take our favourite book from the shelf and start reading. With November approaching what better choice than Surtees and our old friend Mr Jorrocks? We open our volume this evening at the meet at the Cat and Custard Pot.'

The music changes to *Early One Morning*. The lights come up to disclose mine host at the inn with his pretty daughter and other helpers, one of whom is spooning with the local poacher. First to arrive in a gig is the hunt secretary, Mr Fleeceall, then in her

Favourites for a quarter of a century (*Peter Roberts*)

phaeton (*The Surrey With the Fringe on Top* from the band). Mrs Barnington; the first two mounted are Mr Facey Romford of the Heavyshire and Sir Moses Mainchance, Master of the Hit-em and Hold-em-shire (*The Roast Beef of Old England*). In a dog cart, Mrs Jorrocks and her beautiful niece, Belinda (*Here's to the Maiden*), Mr Soapy Sponge and Mr Jawleyford from the Flat Hat country (*Where Did you get that Hat?*); Lucy Glitters, riding side-saddle (*Always True to you Darling, in my Fashion*). Finally Jorrocks himself with little Charlie Stobbs (*Heart of Oak*). The meet is assembled; the whole arena is filled with ponies and riders, all elegant in costume and with the attractive equipages of the drivers.

The second half of the display was the hunt. The lights faded as hounds 'moved off', followed by the field; a clock chimed eleven o'clock, then in the darkness, while the fences were put in position, a record of hounds drawing, then in full cry as one heard the thrilling call of gone away. Suddenly the lights blazed on and headed by the hounds the field raced into the arena and over the fences.

That at any rate was the intention, but it nearly did not happen that way, for when on the Sunday before the Show the cast assembled at Wembley for a rehearsal their ponies just would not jump in the strange surroundings. Colonel Jack Bullen, in charge of the party, tried all his guile, but even if a few could be persuaded to jump the fences the majority stopped or ran out.

Sergeant Major Lee came to the rescue. He had been Mike Ansell's rough riding sergeant in the Inniskillings. After the war he had left the army, but not finding life in industry to his liking he wrote to Mike Ansell and asked if he could work for him. For twenty years, although officially simply in charge of the flower market-garden, he was general factotum and a wonderful help to his old colonel whom he worshipped. From the start Mike had brought him up to Harringay to organise the arena, and so when he saw the trouble with the Pony Club he very quickly reverted to his old role of rough riding sergeant major.

Standing in the centre of the arena, armed with a long-thonged hunting crop, which he cracked with a rifle-shot report, he had all those ponies galloping round the arena and over the fences in no time, with an enthusiasm and gay abandon that many of them can never have known before. It brought the house down as the band burst into *John Peel* (whose centenary it was as, appropriately enough, it was also the centenary of Leach who had been responsible for the original Jorrocks, or Handley Cross, illustrations).

After the ponies had careered out *Just a Song at Twilight* played again, and in the dimmed lights Jorrocks, Pigg and

Charlie Stobbs rode through with the hounds. 'Hunting is the sport of Kings, the image of War without its guilt, and only 25 per cent of its danger,' says the immortal Jorrocks. Black out. A lovely display which lived in the memory for a long time.

The sudden, distressing death of Sergeant Major Lee a few years later was the second tragedy indirectly connected with the Show that was to affect Mike Ansell. For so long Lee had been his right-hand man, but as was his way Mike refused to allow the tragedy to affect him and carried on determinedly.

Among the children taking part in the Jorrocks display were Jenny Bullen (now Loriston-Clarke), her sister Jane, the Olympic gold medallist and brother Charles; Mary and Sheila Barnes, daughters of Gerald and Sylvia, Horse of the Year Show stewards, and both international riders, as was their brother Tom; Paddy Hague one of the outstanding young riders of her time and Rosemary Hayter, who some years later was to become my secretary.

In 1964 the display was revised to coincide with the centenary of Robert Smith Surtees. It was proposed to make it rather more elaborate and I invited a great friend of mine, John Stead, director of studies at the Police College at Bramshill, and an acknowledged authority on Surtees, to help me with the script. It was altogether more ambitious and, on this occasion, all those taking part were adults. Once again it was extremely successful, which was hardly surprising considering the cast list. This included George Hobbs as Jorrocks—surely his greatest Horse of the Year Show ever, for on the last night to the delight of the crowd he won the Victor Ludorum; Douglas Bunn as Mr Puffington; David Barker as Richard Bragg; Derek Kent as Jack Spraggon; Fred Welch as Soapy Sponge; Ronnie Marmont, one of the greatest artists in the show ring, as Pomponious Ego; the immortal Sam Marsh, doyen of show riders and one of the most successful exhibitors before the war at Olympia, where he took part in a famous display, the Midnight Steeplechase, was Lord Ladythorne; Cynthia Haydon was Lady Scattercash; Jenny Bullen, Miss de Glancey; and Harry Goddard, James Pigg with the Enfield Chase hounds, one of whose present masters, Raymond Brooks Ward, was on his feet as Mr Fleeceall.

It is not difficult to imagine the fun and games this cast put into it all, and not surprisingly the audiences loved it though the production had not, in my opinion, quite the charm of the original. The Cat and Custard Pot was revived yet again, with great success, at the Olympia International in 1973.

Many of those who took part in the original Jorrocks display also took part in another very attractive production two years later. This was the Vale of Aylesbury Steeplechase of which, of

Paddy MacMahon and Penwood Forge Mill
(*E. D. Lacey*)

The heavies surge forward at the end of the Musical Drive
(*Peter Roberts*)

The 1970 Ride and Drive Hackney Quadrille. The outrider on the right is the well known National Hunt trainer and rider, David Nicholson (*Davidson*)

course, there is a famous set of pictures making as well known a set of sporting prints as any in the country. The original steeplechase took place in 1834. Among those taking part then were the legendry Captain Becher, of Becher's Brook fame, Jem Mason, the Marquess of Waterford and Mr Allnut, who rode Laurestina and was only eighth at the last brook where all the riders submerged, but was first out and only caught on the post by Vivian ridden by Captain Becher, with Jem Mason third on Prospero.

Living on the edge of the Vale of Aylesbury myself, I was delighted to do all the research and find that so many characters involved such as the Terrys and Peytons still had descendants living in the district. The display again consisted of two scenes; the parade and the race, which necessitated a number of falls to give an air of authenticity. Jem Mason, for instance, in the original race, took a short cut over an enormous rail knowing that his friends had sawn through the top rail the night before. Unfortunately, friends of his rival, Becher, were suspicious, discovered the broken rail and repaired it, thus causing Mason a crashing fall.

Those taking part in the display were Paul Oliver, Alan's brother, and his sister, Vivien, now married to the show jumper, Michael Cresswell; Janet Barnes, Susan Orssich, the daughter of Count Robert Orssich, for so long the producer and rider of many of the most successful hacks in the show ring; and Jane Kidd, John Kidd's sister and daughter of Janet Kidd, who is herself the daughter of Lord Beaverbrook.

The latter, who is now a successful equestrian writer, was involved in quite an amusing incident, though not originally amusing for her perhaps. At one performance she fell off—whether intentionally or not I cannot remember—and was helped from the ring in obvious pain. Before the evening was out her mother was threatening all sorts of terrible things—the Show ought to be sued, Jane should get damages, her horses would be withdrawn. But it turned out to be something of a mountain out of a molehill. Jane, unfortunately, in falling off had had her bottom punctured by a nail. The injury was not very serious and Jane was soon back in the saddle, if a little uncomfortably, but from then onwards every effort was made to ensure that there were no nails or pieces of glass or stones left lying about in the arena. The incident might have been serious though it finished as something of a joke.

To give the race itself an air of reality Raymond Glendenning, the first of the great radio sports commentators, commented on it, making it even more dramatic. It was an exciting and a colourful event and was very popular: but it was never easy to give it the same nostalgic atmosphere of The Cat and Custard Pot.

The revival of the Jorrocks display in 1964 was really the last of the big tableaux, though throughout the existence of the Show there have been, until the last few years, some very high-class high school displays. These started with M. Krier's first appearance in 1949, and then the Olympic gold medallist St Cyr of Sweden in 1951. One of the most impressive was the quadrille put on by Baroness Ida von Nagel's Vornholz Stud. Accompanying her were Otto Lorke, at one time the Kaiser's riding master, a previous pupil of Lorke's, Willi Schultheis, at that time Germany's leading dressage rider and the outstandingly versatile horseman, Fritz Theidemann who won bronze medals in the Olympics both in dressage and in show jumping. In the latter his most famous horse was Meteor. All the horses taking part were bred at the Vornholz Stud, their quadrille being based on the classical exhibition which could be traced back 400 years.

It was in 1953 that the display of classic riding that most caught the public's fancy was first performed. This was the superb dressage exhibition by Madame Lise Hartel, the Danish Olympic silver medallist at Helsinki. What was so remarkable was that she suffered from polio and had to be lifted into the saddle. She was a most beautiful and sympathetic horsewoman, and such was her success at Harringay in 1953 that she was invited back in 1955, by which time, in addition to her Olympic silver medal she had won the Danish championships three years running, the Nordic Games twice and the FEI Grand Prix in 1954. Her horse, Jubilee, by Rockwood, was a most impressive mare and the performance by the two of them was unforgettable.

The display in 1966 was very different. In the programme it was described as 'dressage at the gallop' and 'a breath-taking display of advanced haute école, carried out, of necessity, at very great speed', performed by 'one of the finest horsemen in the world, a great showman, whose amazing dash and skill make a remarkable spectacle'. All very exciting: yet the display caused a considerable furore. Señor Don Angel Peralta was at that time, with his brother, the leading rejoneador in Spain, a mounted bull fighter. At one or two performances he did give demonstrations of how the Andalusian horses—in his case Rusenor and Faroan—were trained to a bull by using a man running with an artificial bull's head on a kind of one-wheeled bicycle, but this display was primarily an exhibition of brilliant horsemanship—quite literally, dressage at the gallop. Not surprisingly, those opposed to bull-fighting felt that Peralta should never have been invited as it implied the Show's approval of bullfighting.

This led to an incident which was not without its amusing side. That very remarkable person, Mrs Glenda Spooner, founder and chairman of the Ponies of Britian Club, and a tower of strength

Joe Moore flat out in the Eldonian Double Harness scurry (*Lane*)

in many fields connected with the welfare of animals, wrote to Mike Ansell complaining about the display. He replied giving his reasons as to why he thought it was justified, wording the letter strongly, but not offensively. In effect the implication was—you run your show, and let me run mine. Mrs Spooner was not satisfied with this response and wrote back briefly saying that she had no option but to accept the situation, but that she thought that he was wrong. To a friend she wrote in terms a good deal stronger. She considered Mike Ansell to be a tyrant, a bully, that he would never listen to anyone else's opinion, had to have his own way etc.

A week or so later Glenda Spooner was surprised but gratified to receive an invitation from Mike Ansell to call upon him at Bedford Square, which was where the Show office was then situated. She arrived: he offered her a drink which she declined. He asked her to be seated, in the chair across the desk from him.

'I received your letter,' he started, and pushed across the desk the letter that he had in front of him. She picked it up, realising at once that this was the letter that she had written to her friend. They had been put in the wrong envelopes. For a moment there was silence, then she barked out that rather mirthless laugh which anyone who is familiar with Glenda on the war path knows only too well:

'So now you know what I really think,' she said.

Mike Ansell was at his most charming. It was a case of Greek meets Greek. Neither was likely to be persuaded to the other's point of view, but at least they respected each other.

Willie Schultheis came back in 1961, this time to Wembley, and with Frau Springer, an eminent German lady rider. Their Pas de Deux was particularly popular and many had the opportunity of seeing for the first time the perfect harmony possible when two horses and riders are perfectly trained. Another brilliant 'double act' was presented when the Swiss riders Henri Chammartin and Gustav Fischer came over in 1965: the former won the Olympic gold medal in 1964, having five times won the European Dressage Championship; the latter was fourth in the 1964 Games and won the silver medal in 1952.

Their invitation could have been a compensatory gesture by the show director for his determination to assert his independence the previous year by inviting Peralta back. In fact, this had then caused little controversy and less than the display the following year by the Portugese maestro, Nuno Oliveira. Riding, alternatively, two very impressive bay Lusitanian stallions, in eighteenth century costume, Oliveira gave a highly entertaining display. The great majority of the audience enjoyed it enormously—though not the purists who felt that it was a circus act, inaccurate and indicative of less than thorough schooling. Indeed, they pointed to the fact that the more dramatic of his two horses, Curioso, was only five years old and could not possibly have attained grand prix standards by proper training methods: three or four years at least are reckoned to be needed to bring a horse to grand prix level.

A similar criticism was levelled at the Dutch rider, Henry Gilhuys, when he came over with a four year old, as well as his twelve year old, Deauvillois, in 1969, and again in 1971. It emphasises, on two levels, two different lines of thought. There are those who prefer the disciplined German style of dressage, demanding complete accuracy: others prefer the more fluent French style. Again there are those who feel that the emphasis in a display at the Horse of the Year Show should be entertainment value; while others feel that any display of advanced equitation should be fundamentally correct: that a display is an example which others will try to follow. Most, I think, would agree that on the whole the latter attitude is unnecessarily rigid. The Horse of the Year Show has always been an entertainment, a spectacle as well as a horse show.

Mention of the brief but effective appearance of the Spanish Riding School on the last two days of the Show in 1969 should not be omitted. The next year saw a very different, though equally effective display by the hackneys. Indeed, many thought it to be one of the best of all the Horse of the Year Show displays. There were four drivers, led by Cynthia Haydon and four riders who included David Nicholson, the trainer, David Barker, the European Show Jumping Champion in 1962, Dick Stillwell, the trainer of so many top-class jumpers and eventers, and Jane McHugh, a leading show rider.

Nineteenth century costume was worn and immense trouble was taken with the music and the lighting; but, perhaps, what really created the atmosphere was the placing in the arena of six old-fashioned Victorian street lamps. At the beginning of the display the arena was in darkness, then with the opening announcement and the evocative, 'trotting' music the lamps came on, and in drove the carriages with their own lamps glowing, each vehicle escorted by a mounted rider. The great success of this display was due to the Haydons whose reputation with hackneys is international, and Bob Black, Cynthia's brother. They are all great characters as well as Hackney experts and can be guaranteed to make any party go, inside or outside the arena. From the Show's point of view it is their complete reliability and professionalism that has made such an important contribution, especially where anything to do with harness is concerned.

In *Soldier On*, Mike Ansell quotes Prince Philip who said to him on one occasion that he was tired of watching young people

The Riding Club Quadrille; an elegant interlude: the 1975 winners, East Grinstead (*Lane*)

The Pony Club Games: Enfield Chase, winners in 1961, in one of the exciting races (*Lane*)

trotting and cantering around the ring on their show ponies, obviously getting into trouble with their parents if they did not win. Would it not be possible to arrange races or something for them, riding ponies that need not necessarily be show ponies and making it more exciting for the audience? 'I was delighted', Mike writes, 'that was just what we wanted, so we set about organising the mounted games: for teams of four, no prize money, only rosettes.' Displays, he felt, were getting more and more difficult to find, and costing more and more. To replace them with the Pony Club games would be both a saving and a popular entertainment.

That the mounted games have been popular none can deny: nor that in present conditions it is essential for the Show to make as much money as it possibly can, but there are many, I believe, who look back with a certain nostalgia to the displays of the earlier days and regret their disappearance. However, to discontinue the mounted games would be even more regrettable, for in a very short time they established themselves as one of the most popular items of all, creating an enthusiasm in the audience that has often been likened to the enthusiasm at a cup final. Indeed, with the cup held aloft by the winning team in their lap of honour, and ribbons in the team's colours streaming from its handles, there is a strong similarity to the event that takes place six months later in another stadium barely 100 yards away.

The Prince Philip Cup was first held, experimentally, in 1957, at Harringay, when teams in the vicinity of London were invited to compete. The following year it was open to teams from branches all over the country. In 1958 no less than 164 branches entered, taking part in 16 regional competitions. From these, 33 teams qualified for six zone finals, the winning team from each zone qualifying for the Horse of the Year Show. Each team consisted of five riders, but only four represented the team in any one event. At each performance there are four events and points are awarded for each race, 4 to the winner, 3 to the second and so on. At the end of the week the four teams with the greatest number of points compete in the final.

The atmosphere is electric from the first moment that they ride in to the strains of *Boys and Girls Come Out to Play*, and parade round the arena, each team dressed in jodhpurs, a different coloured shirt with a Pony Club tie, and a velvet cap. They all file out, returning immediately for the first race, usually 'bending'. Immediately the cheering starts, led by the team supporters and those in the teams not competing who are seated in the exhibitors' seats behind the band. Everyone shouts encouragement for his own team. Before the first race is over the whole audience joins in, including the VIP's in the royal box. 'Blue! blue!' 'Come on

*red*!' or the actual name of the team, 'Enfield Chase!' 'Strathblane!' Raymond Brooks-Ward who has made a speciality of both running the games and commentating on them, leads and provokes the cheering, yet he can hardly be heard because of the crescendo of din running right through each game. '*Green*! Come on *Green*!'

Different games are selected for each performance, some of the most popular being the sack race, the stepping stone race, the Daily Mail race (the *Daily Mail* sponsors the event) in which riders have to push a newspaper through a letter box, pick up a penny (?) hand it to the next competitor at the end of the arena and so on; and the sharp-shooting race in which the riders gallop down the arena, dismount and have to knock the head off a plaster figure.

The expertise and the skill of the young riders is incredible, and again and again they perform marvellously acrobatic feats. They are able to vault onto their ponies from any position and if nothing else they learn tremendous team spirit from the games.

Inevitably there are people who say they make children ride in a rough-and-ready style. Obviously this is true in isolated cases, but indeed one would not expect refinement, and the benefit of learning to ride with balance, suppleness and dash must outweigh any disadvantage. Above all it makes a wonderful spectacle and is a fine advertisement for the Pony Club. The cheer as the children finally exit, the be-ribboned cup held aloft, all but brings down the roof, drowning the band playing *Kids*. In nearly twenty years some 200 teams have taken part since the North West Kent team won the first games in 1957. Both the High Peak North and the Atherstone teams have won three times and for Scotland, the Angus and Strathblane teams have been successful.

It is right that children should be prominently associated with the Horse of the Year Show for it is so essentially a show that is young in heart. There is plenty to be taken seriously but there is, too, plenty of fun and games.

A superb capriole on the occasion of the visit in 1969 of the Spanish Riding School of Vienna (*Roberts*)

# 11 Jumping and TV

*The chicken or the egg—reasons for its popularity—all that matters on the screen—Helsinki success—a complicated technique—homework—the mental strain—producer-commentator relationship—the early days of television—political brinksmanship—new sophistications—a driving mishap—the annual 'must'—the first* Grandstand

So often one hears people say that television has made show jumping—and I have heard a few people say that show jumping has made television! It is undeniable that television brought show jumping to millions of people who would never otherwise have been interested in a sport which tended in earlier times to be exclusive. It is an ideal sport for television coverage: first because it is exciting and pictorially effective, more so than ever since it has been televised in colour; secondly the rules are simple, which makes it easy for a layman to understand. Though obviously there are certain technicalities, for the most part it is just a matter of a horse clearing a fence or not clearing it, and then in a jump-off the fastest horse winning. One does not have to be an equestrian to understand something as simple as that. Many sports, because of their abstract rules, are too complicated wholly to involve the lay viewer.

There is an even more important reason for the success of show jumping on television. It is a fact that those sports of which the vital ingredients can be contained on the little square screen are the most popular. With few sports can this be more the case than show jumping. All that matters is the fence, the horse and the rider and these can be clearly in view all the time, with the result that watching becomes compulsive. Sports like tennis, gymnastics and boxing are popular for the same reason: in tennis the camera can take in at the same time the ball and the two players; in gymnastics there is only one performer upon whom the cameras can concentrate without distraction; in boxing the two protaganists are visible all the time and so hold the viewers' attention. In football, for instance, if a player in the centre passes out to the wing one cannot be sure that a player is there waiting to receive the ball until the camera finds him. In cricket, a batsman can hit a ball to the boundary, but again until the camera has found him the viewer is unaware whether there is a fielder there or not. In

The High Peak team, winners of the Pony Club Games in 1975 (*Findlay Davidson*)

Caroline Bradley on Marius (*Peter Roberts*)

golf, the ball is frequently lost sight of altogether. In horse racing it is invariably difficult for the camera to cover the whole field, as is the case in sports like motor racing, motor cycling, bicycling or swimming.

So show jumping has a great advantage, as the viewing public was quick to appreciate. Even in 1956 I can recall an announcer in a programme of that time entitled *Programme Parade* saying: 'perhaps the most popular television programme of the year—the Horse of the Year Show'. I have always believed that it was the Finality/Foxhunter duel in 1950, described in the next chapter, that first alerted the viewing public to the fact that here was a new sport that could be entertaining, even compelling. Consolidation came when in 1952 the British team won the show jumping gold medal in the Helsinki Olympic Games. As this was the only medal won by Britain in those games and as it was only at the eleventh hour that it was won (show jumping is traditionally the final event before the closing ceremony), enormous publicity was given to our victorious riders, Colonel Harry Llewellyn, Colonel Duggie Stewart and Wilf White—and to the sport they represented.

Show jumping at that time was a novelty as far as the British public was concerned. The success of our team that July in Helsinki made thousands flock to Harringay a few weeks later; made millions switch on to watch the Horse of the Year Show on television. It was at about this time that the head of a well known brewery told me that his publicans could always tell when the Horse of the Year Show was on because suddenly the bars were emptied. Later commercial television was to discover that when the Horse of the Year Show was on they could lose as many as $1\frac{1}{2}$ million viewers in an evening, a fact which seriously affected the letting of their advertising space.

Today the Horse of the Year Show attracts between 65 million and 70 million viewers during the week, with viewing figures for each evening varying from 9 million to 13 million. In a survey published in *The Field* a few years ago it was stated that show jumping was the second most popular television sport, second only to football.

Not surprisingly the viewing public have very little idea of the complicated background to a television broadcast. Months in advance the producer meets the Show organisers to discuss transmission times and each programme has to be slotted into a plan. A certain amount of give and take is therefore necessary. For instance, the Show may normally be planned to finish at 10.45 pm, but television may want to start another programme at 10.30 pm. Either the Show must bring its own programme forward, or the television authorities must agree to put back its next programme until 10.45 pm. Alternatively because of, say, a

Harvey Smith on Salvador (*Findlay Davidson*)

The famous Cavalcade, the Show finale (*Findlay Davidson*)

The most atmospheric arena in the world (*Stonex*)

party political broadcast, it may not be convenient to start transmission of the Show until 10.15 pm. This would mean that from the television point of view, the event covered should end at 11.00 pm, or even 11.15 pm although this would be considered too late for the Show. Unless, therefore, an exception is made, delaying the end of the Show until 11.00 pm, it will be agreed to fill in the last 15 minutes of television time with part of another competition recorded earlier.

It may take a certain amount of to-ing and fro-ing between the BBC and the show office at Belgrave Square, before the transmission times are finalised. The producer can then start planning in detail. How many cameras will be needed? Where will they be placed? This is crucial for if, as almost certainly will be the case, cameras are needed other than at the ringside, the BBC must pay for the seats displaced.

Will a camera be needed outside the arena? In the collecting ring, for instance? or even in the conference room for interviews? Permission will have to be sought from the Wembley authorities. Where will the commentary position be? More displacement of seats: more negotiation. Lighting: how much extra lighting will be required? What are the technical problems involved? Where can the scanner be situated? Is a portable camera required? If so where can it be positioned?

The rigging has to be planned, again requiring the co-operation of the Wembley authorities, the personnel briefed, the unions consulted, commentator, link-man, captions, stadium manager all lined up. The producer is fully occupied for many weeks before the Show, and can leave nothing to chance.

Nor is it quite as simple for the commentator as viewers often think. Many people have told me how they envy me my job, just sitting at the ringside and chatting into a microphone when I feel like it. Apart from the homework that has to be done to be familiar with the horses and riders, and conditions of the various competitions concerned—like the tip of the iceberg only a fraction of this information is ever used—there is the concentration. It is not possible to relax for a moment. Through earphones I listen, all the time, at least with one ear, to the producer who has to be in contact with many other people besides oneself: the cameramen, the captions department, the studio manager, possibly even to 'presentation' back in London. Through it all I have to have one ear open all the time for anything that concerns the commentator. For instance the producer, acting as Mr Everyman, can often suggest to the commentator a description or an explanation or an identification which will help or interest the viewer.

A really lengthy session such as the Burghley or Badminton

Friends and rivals whose duels over the years have drawn and fascinated the crowds: David Broome and (over page) Harvey Smith (*Stonex*)

Harvey Smith (*Stonex*)

Horse Trials, or a weekday afternoon at Hickstead, or some such important occasion as the Olympics, or World or European Championships, can be very exhausting, though not, as people often imagine, for the voice. This in fact is seldom strained as for the most part the commentator is talking quietly into a microphone. It is the concentration, rather, which is mentally exhausting.

As far as the producer is concerned, the commentator is simply a cog in the very sizeable wheel that he is turning. Sometimes it is annoying if a producer tells you, for instance, to say how many clear rounds there have been. In fact, you may have just given out this information a few minutes earlier, but at that moment he was talking to someone else or was in some way or other distracted. Naturally you cannot explain this to him because anything you say is relayed to a million homes, so you can but sit and suffer in silence! But as far as the producer is concerned, the commentator is but one of his responsibilities. Understandably he likes commentators who are experienced, about whom he need not bother himself very much. Conversely a commentator likes a producer who is experienced and who leaves him alone without too much prompting: as much of his concentration as possible can then be given to the event he is covering, rather than listening to his earphones.

Naturally, the television coverage of show jumping today is far more technically sophisticated than it was in the early days. At times I feel that with television, as with so much else, this can lead to a situation where you can scarcely see the wood for the trees, the raison d'être for the transmission being lost in a welter of technical and scientific brilliance. It was certainly much simpler in the old days. Today producers are extremely knowledgeable about the sport they are covering: but it was not always so. When show jumping was first televised the producer had to rely almost entirely on his commentator to help him to find out what it was all about. Yet there was something rather enjoyable and relaxed about the informality of it all, as is perhaps exemplified by my own introduction to television.

Peter Dimmock, then head of outside broadcasts, was in the audience at the first Show. At the end he leaned over the balcony rail in front of the seats above the commentary position and called out to me: would I be interested in doing television commentaries? When I told him that I thought I would, he promised to get in touch with me. My premier obligation at that time was, I told him later, to the public address which I could not desert. Since I had started at the International Horse Show at the White City in 1947, I had been trying to develop a different technique for it; informative and lighter than the rather solid, strictly factual older style. I suggested that I could quite easily manage both the public address and the television commentary, using two microphones. This idea was happily accepted. Strangely, in some ways I found it easier doing both commentaries; one was complementary to the other; there was no fear of repetition, or contradiction—no cutting in.

Later the BBC found this technique unsatisfactory; so, possibly, did the Show. There was the danger of serving two masters, each of whom tended to feel that the other was getting the better deal: but in those early years it seemed to work. The first show jumping producer, Bill Duncalf—later to achieve great success with his medical programmes and his series on gardens—appeared to be completely happy with the method. Indeed, he liked me to interview show jumping personalities as well. I can remember a number of occasions when in the short break between the first round and the jump-off, while the fences were being raised, having given out the clear rounds and the height of the new fences, I would then leap out of my commentary position, double down the back alley to interview one of the finalists, then race back, just in time to announce the start of the jump-off. Very occasionally I would pick up the wrong microphone, but no one ever seemed to worry unduly. (It may be a surprising fact that one can speak more intimately to ten million unseen viewers than to ten thousand listening to the public address in the arena.)

Nor did the BBC worry so much about an over-run as it does today. In those days, the timing of a competition, the building of a course to produce the exact number of clear rounds desired, had not become the fine art that it is today. Frequently, therefore, a competition would over-run by as much as thirty minutes. Obviously the programme that followed and the viewers awaiting it were not very pleased, but they certainly seemed to prefer this to being cut off in the middle of a competition.

I remember that on one occasion the annual conference of one of the major political parties was being held at Brighton. About twenty minutes before the scheduled end of our transmission it became obvious that we were going to over-run by a quarter of an hour. I informed my producer of this on the device known as the lazy talk-back. Not surprisingly, he told me that there was nothing that could be done about it as the political conferences were sacrosanct, an over-run would never be accepted: we would just have to miss the jump-off and have the result announced later, a ploy which I knew would infuriate viewers. I could understand well enough the BBC's point of view: it was not their fault that we were running late. I therefore resigned myself to the unsatisfactory situation of a programme cut off in its prime, knowing that I would be bombarded with complaints from viewers. A few minutes later, to my great surprise, the producer told me through my earphones that much to his amazement 'presentation' had informed him that they would take the jumping to the end of the competition. They had been told by the authorities that under no circumstances was the programme to be curtailed—the party concerned reckoned that to cut off the jumping before the end would cost them a million votes!

Today immense care is taken to see that a programme does not over-run. Very occasionally due to an unexpected number of clear rounds it happens, but it is more likely to be the other way round. An under-run is much easier to provide for as it is always possible to broadcast a jump-off of an earlier competition.

Transmission today is very sophisticated—too sophisticated, perhaps. Whereas in the past when the time came for our programme the announcer back in London just used to hand over to 'our commentator, Dorian Williams', and let him get on with it. Now there is the very effective Horse Show signature tune *Waldos los Rios*, Mozart's musical joke, and a montage of action shots which form a background to the captions, then over to the link-man, David Vine, to 'set up' the programme, and finally, only inside the arena does the commentator take over. At the end of the round a leading rider is dragooned to submit to a brief interview while the jump-off course is being prepared. If no rider is available a few minutes of another competition may be inserted before returning to the arena. At the end, rather than show the presentation of prizes which can be a somewhat protracted affair, though many viewers appear to enjoy it, there will either be another interview or the jump-off from an earlier competition, or possibly, a few minutes from one of the displays, or the Pony Club games, or even the musical drive of the heavy horses.

The value of all this, of course, is the variety. In an hour of viewing it is possible to show the highlights of two major jumping competitions, part of a display, and meet face to face, as it were, one or probably more interesting personalities. In addition it is to the advantage of the Show to be broadcast on television, so encouraging the contributions of its sponsors.

It is perhaps arguable that too much variety detracts from the status of a major event. It is not entirely inapposite to liken it to the men's final at Wimbledon sharing space or even being interrupted by a lesser event. I cannot deny that as a commentator I found it easier to build up a competition to a real climax when there was an uninterrupted run from the moment the programme started right through to the end—even the delay of raising the fences could be made part of the build-up. My technique was to set the scene in the first few minutes, identifying competitors and explaining the course and its problems, then to say less and less as the competition developed until, ultimately, no words were required at all: an 'ooh!' or an 'ah!' sufficed. This technique is not so effective in an interrupted, or recorded programme, for obviously, the lack of continuity makes it essential to 're-cap', although probably to the annoyance of viewers who have watched throughout.

But this is a personal opinion. The likelihood is that the majority of viewers prefer the variety, prefer the cleverly presented package programme, rather than the one straightforward transmission. Certainly it is to the benefit of the Show, both because of the increased revenue and because of the opportunity for the viewing public to see that the Show is much more than one major event—as with a football match, for instance—thus enticing them to come to see for themselves.

Variety, too, is the hallmark of the good producer, even at the Horse of the Year Show: cut-away shots of the crowd, the royal box, a glimpse of the collecting ring, or the band, or even control; the smart young men in the arena party, the course builders, the clock superimposed in a jump-off, a course plan, flags, the trophy. All these tactfully, almost surreptitiously, filtered into the programme enormously heighten the interest. I doubt if any interpolated shot has been more effective than the little trap turning over in a driving event, the very large but sporting lady driving it then picking up her skirts and chasing after the ponies. This incident, it is probably true to say, brought one of the biggest cheers ever heard at Wembley. It is not surprising that it was frequently used by the BBC.

Of all the producers with whom over twenty-five years I have been privileged to work the most talented was without doubt Alan Mouncer, who left the BBC in 1974. No one produced more effective television than he, yet his successor Fred Viner who worked with him for a number of years, may with his rather simpler method of presentation, serve the actual sport of show jumping better. What is certain is that it is the skill of those producers who for nearly thirty years have been presenting the Horse of the Year Show on television that has made show jumping one of the few sports that has survived, indeed benefitted from the television exposure to the mass audience. The way that it has been presented has built up a huge television audience and made people want to see for themselves: thousands have thus been lured to Wembley and found that there is so much more to it all than they had ever seen on television—however skilfully produced, television can never quite convey the remarkable atmosphere or variety of the Horse of the Year Show. Having come once, they come again—and again and again.

In 1975, BBC's *Grandstand* celebrated its 1000th transmission.

It included show jumping from the Horse of the Year Show, as it did on the first transmission in 1958. On each occasion I was commentating. Before the first programme I received a telegram from Paul Fox, the head of outside broadcasts: it read, 'DELIGHTED HAVE YOU FIRST GRANDSTAND PROGRAMME STOP APPRECIATE THERE ARE CHANGES FROM NORMAL ROUTINE BUT SINCERELY HOPE YOUR EFFORTS WILL MAKE THIS FIRST PROGRAMME A SUCCESS STOP THANKS IN ADVANCE YOUR CO-OPERATION STOP REGARDS PAUL FOX'.

I doubt if television viewers have been aware of any changes. It is still, for millions, a 'must', just as for thousands more, who come to the Show in person, the annual visit is a 'must'.

# 12 Magic Moments

*Different types of competition—past heroes—Alan Oliver versus Seamus Hayes—Ray Howe's Puissance—George Jorrocks Hobbs—David Broome and Sunsalve—Paddy MacMahon and Penwood Forge Mill—comic cuts with Nugget—Foxhunter and Finality's classic duel—Schockemohle versus Broome—Broome's clean sweep*

The great majority of television viewers have little idea of the variety that comprises the Horse of the Year Show. For them it is just a series of major jumping events. It is true that although the Horse of the Year Show is a spectacle with many and varied ingredients, it is principally a jumping show. Certainly it is top-class jumping that has dominated the Show down the years: and there have been many magic moments.

The pattern, ever since the beginning, is more or less regular. At the beginning of each performance there is a jumping competition, open only to top-class horses and riders, but of less stature than the major event that always occupies the whole of the second half of the programme. The earlier competition is frequently a speed event, or one that is a little unusual such as a 'touch and out', a 'gamblers' or an 'accumulator'. These competitions add variety to the performance and being, as a rule, easy to time make it easier to organise the whole programme so that it does not run late. The importance of this lies in the fact that the last jumping competition is invariably televised, and naturally the television likes to keep to exact timing.

Although the Show usually opens with a lesser competition it is frequently as popular, or even more popular, with the audience than the major event. One can recall from the earlier days some wonderful speed events in which such horses as Monty, Niblick, Planet, Costa, and Go Lightly, gave the crowds some tremendous thrills. Later we were to have Prince Hal, Craven A, Red Knight, Sheila.

One of the most exciting jump-offs that I can remember happened in 1952. It was in the aptly named Gordon Richard Stakes, in which the premium was very much on speed. Seamus Hayes had won it the year before on Mr Tommy Makin's Planet, beating Alan Oliver on one of Mr Payne's horses. Now it was Alan Oliver again, this time riding Red Star. Alan appeared to have set Seamus a really stiff time to beat, but Seamus came in with all his usual abandon—there were few, if any, more popular riders with the crowd at that time—and he fairly hurled himself round the arena. With two fences to go he appeared to be fractionally up on time, but as Planet landed over the second element of the double of railway gates, he tripped. Within a couple of strides he was on his knees with his nose on the ground. It seemed as though he must, at the next stride, topple over. But somehow Seamus snatched him up, got him into a stride, miraculously balanced him, and forced him to take off for the last fence, another double. He raced home a winner by one fifth of a second.

It was an outstanding exhibition of horsemanship, as was the occasion when he won the Puissance on a big grey horse, Doneraile, that he later sold to the Italians. Coming into the wall, then standing at over 7ft, his horse took off a stride too early, pulling the reins out of Seamus' hands. Somehow he retained his contact and though as the horse sailed over the wall Seamus' hands were high above his head he maintained his balance, seizing hold of the reins again as he landed.

Sir Malcolm Sargent presents the Sunday Graphic Cup, the Victor Ludorum, to David Broome on Sunsalve (*Lane*)

Another unforgettable Puissance was won by Ray Howe in 1972 when, the only one to clear the 7ft wall, as he landed he took off his hat and threw it right up into the roof: a wonderfully effective moment on television.

The Victor Ludorum on the last night has provided some memorable competitions. There is already a unique atmosphere that night: a popular victory in a really dramatic jump-off ignites the whole stadium. In the previous chapter I mentioned the 1964 Jorrocks display, the name part being taken by George Hobbs. Within an hour or so of his personal triumph as John Jorrocks George was back in the arena on his consistent international ride, Royal Lord, and to the delight of the packed audience he won a really good Ronson Trophy, as the Victor Ludorum was then called. It was somehow a case of poetic justice, Hobbs having been so long a stalwart of show jumping yet never quite reaching the uppermost heights though he has been second in the King George V Cup no less than four times. This is the sort of situation that the Horse of the Year Show crowd relish and appreciate. That night George Hobbs probably had the greatest reception of his career.

No one who was present will ever forget David Broome's triumph on Sunsalve in 1960 when he had earlier that year won the bronze medal in the Olympic Games. As he turned for the last three fences, an extraordinary crescendo of cheering started, mounting to a mighty climax as he cleared the last fence. It was as though this great cheer literally lifted him over the last three fences, the crowd suddenly becoming aware that it was possible for him to beat the time, then willing him on and over. His partnership with Sunsalve was the most popular in his career.

This sort of crowd involvement was even more noticeable in 1972. Coming only a few weeks after the Munich Olympics, and immediately before the very valuable Courvoisier Championship with prize money worth £20,000, the Horse of the Year Show attracted a particularly strong overseas entry, spearheaded by the Germans and Hartwig Steenken with his great mare, Simona. As so often on previous occasions the Show developed into a duel between the Germans and the British. It was in the Victor Ludorum on the last night that it reached an unforgettable climax. The Victor Ludorum is a two round competition. The best third from the first round goes through to the second round. Those with equal faults in the two rounds compete in a jump-off against the clock. On this occasion there were eight with double clear rounds: three from Germany, four from Britain and young Eddie Macken from Ireland, who made such a remarkable impression at his first international show.

Steenken and Simona were the first to go and went clear in a fast time. No one else went clear until the third German rider, Paul Schockemohle, Alwyn's brother, on Agadir, whose time was faster than Steenken's. Last to go was Paddy MacMahon with Penwood Forge Mill. Although he had narrowly missed the Olympic team Forge Mill was certainly not the famous horse he was to become the following year when he won the European Championship and the King George V Cup within the space of one week in July. Two nights earlier he had only just failed in the Puissance, so effectively won by Ray Howe on Kalkallo Prince, tying for second place with Hartwig Steenken on Der Lord. A few weeks earlier he had gone very close to winning the Hickstead Derby. But that was in the wide open spaces of Hickstead. Forge Mill is a big horse, needing plenty of room. How could he cope with the confined space of the indoor arena at Wembley when trying to beat the clock?

Yet somehow it was apparent to everyone from the very moment he came in, saluted the judges, and cantered round awaiting the bell, that Paddy was out to win. As he rose to the first fence the cheering started, and as with David Broome ten years earlier, it increased in a deafening crescendo as he jumped each fence. As he turned for home, the crowd, stealing a glance at the electric clock slung over the centre of the arena, realised almost instinctively that he was in with a winning chance. Not only did the cheer rise to a roar but the huge audience leapt to its feet as one, to shout him home. He won by a tenth of a second, notching up for Britain at the eleventh hour a much-needed victory over the Germans on home ground.

Paddy MacMahon received a hero's welcome when he returned for the presentation: a reception that was repeated when he appeared a few minutes later in the cavalcade. Again it was the sort of occasion that seems to be part of the history of the Horse of the Year Show: pent up excitement, suspense, triumph. And humour. There have certainly been plenty of laughs over the years. There has never been any shortage of extrovert characters among the show jumpers, ready to seize upon any opportunity for a bit of clowning—and not only in fancy dress or bare back competitions.

The funniest occasion I ever remember was at Harringay in 1956. It was a 'gamblers competition' in which each of the twelve fences in the arena had a different playing card, but only seven of which had to be jumped. The most difficult of all was a very big treble down the middle which was accredited with the Ace, worth 13 points. At that time a rider could jump one fence as often as he liked, but not more than seven. What followed was so complicated that I feel I cannot do better than quote from an account that I wrote of the event at the time.

George Hobbs wins the Victor Ludorum, then the Ronson Trophy, on Royal Lord in 1964, the year he featured as Mr Jorrocks in 'Jorrocks rides again' (*Roberts*)

The two protagonists were Seamus Hayes on a grand Irish-bred hunter type called Waving Corn and John Walmsley on Nugget, a Welsh Cob and a great personality; also, a very good show jumper. At Harewood in Yorkshire he won the Ladies Championship ridden by Anne Morley, who later became the wife of John Walmsley, who then proceeded to win the three other major competitions on him.

When Nugget came in nobody had scored the maximum, which was, of course, seven times thirteen. John Walmsley wasted no time and went straight for the big treble. Down the line he went, faultlessly, John riding on a completely loose rein. He turned at the end of the arena, and with three or four strides off he went again, clearing the treble a second time. A turn at the end and he attacked it for the third time. So he continued, sometimes getting so close to the second or third element that it seemed as though he took about six strides between the two elements, rather than the two which had been anticipated by the course builder. But then, at about the fifth attempt, he hit the third part of the treble and down it came. So he jumped the King —for twelve points—and then, the pole having been put up again, back he went to the treble!

By this time everyone was in hysterics. They had completely lost count, and the style of Nugget's jumping not only had the packed arena at Harringay in fits of laughter but several million viewers at home: and to be honest, the commentator as well. Eventually Nugget had jumped his seven fences and retired with a score of seventy-seven.

The next rider to come in was Seamus Hayes, on Waving Corn. He knew that the rules included one to the effect that if a competitor knocked down part of a combination, he then had to go back and jump the whole thing again, and in this competition it was only a fence cleared that amassed points; there were no faults for a refusal or a fall. He therefore employed the same tactics as Nugget, but whenever he hit a part of the combination he made his horse stop at the next part, dashed back to the start, held up his hand to show the judges that he was ready to begin as soon as the fence was re-built, and they had rung the bell. In this way his seven fences seemed endless, but he carried on, frequently faulting at the first or second part of the treble, and then stopping at the second or third, and so collecting no faults, but still collecting a maximum of thirteen each time he actually jumped the complete treble.

This went on and on for minutes, the laughter even greater than it had been for the bouncing, short-tailed Nugget. I cannot remember any serious show jumping event at an important show getting such a hilarious reception—except, perhaps, when Harvey Smith took the bareback riding competition at the Royal International Horse Show in 1972 a little too literally and actually removed his shirt! At last Seamus Hayes finished and, to roars of applause and cascades of laughter, he rode out of the arena waving cheerfully to the crowd, but just as he reached the exit the score was announced. There was a sudden silence as the audience listened to it. But what an anti-climax. Seamus had been disqualified. Apparently he himself had lost count of the number of times that he had cleared the fence and so, when the judges added it all up, they found that he had jumped the fence once too often. The judges had no option but to disqualify him. Seamus' reaction, still playing to the crowd, brought the house down. For the record it was old Nizefela who eventually won the competition. Fortunately, thanks to the changes in the rules, nothing quite so complicated can ever again happen.

But most of all the many highlights in the show jumping at the Horse of the Year Show I still believe that it was the first real duel that ever took place at the Show that was to have the greatest impact of all. This was the puissance at Harringay in 1950 when the competition resolved itself into a David and Goliath battle over two huge fences. Pat Smythe on her little Finality, the 15 hh mare that she was riding for Mr Snodgrass to whom the mare had been sold as described in an earlier chapter, represented David, while Goliath was none other than Colonel Harry Llewellyn, with his mighty Foxhunter, fresh from winning the King George V Gold Cup.

Again I recorded my impressions of the unforgettable ding-dong battle immediately after the event. Later it was published in *Great Moments of Sport: Show Jumping*. I do not think I can improve on this description of an occasion which I have always believed was to have a vital influence on show jumping in this country, simply because it was the first show jumping event ever to show how ideal show jumping was as a sport for television.

Finality had to jump first. The wall stood at over 6ft. The first fence was a huge triple bar with a spread of nearly 7ft. Finality jumped this awkwardly, but carefully. She turned at the end of the arena, sailed down to the wall and cleared that too.

It was then Foxhunter's turn. He jumped first the triple bar and then the wall—both clear. Both fences were then raised. Little Finality came in again. She met the triple awkwardly again and, although she made a great jump, screwing over it, she had a pole down. She then turned to jump the big wall and, again, only just got over this: 4 faults. This seemed to be handing it to Foxhunter. He sailed down to the triple and jumped as though it were nothing at all, but turning to the wall he appeared to be a little over-confident, jumped it carelessly and, just tapping a brick, it came down, to give him 4 faults, and so they were equal once again.

The fences were raised. There was an absolute hush as the diminutive Finality and the youthful Pat Smythe rode into the ring. She galloped down to the triple bar, screwed over it, landing well to the left hand side, but she was clear. Then Pat did a truly remarkable thing. She pulled up Finality altogether, many people thinking that she had decided that the big, red wall was too much and that she was therefore going to retire. But not at all, she was just giving her little mare a brief opportunity to get properly balanced. She turned at the end of the arena to face the wall, gathered the reins, applied the pressure of her legs to Finality's sides and set off towards this great wall. She only cleared it by an inch or two, just rapping it with the hind hooves, but she was clear. To keep in the competition, the great Foxhunter had to go clear again as well.

He came in looking every inch the greatest show jumper in the world—as undoubtedly he then was. He jumped the triple bar safely, and this time Colonel Harry Llewellyn was allowing

Ted Williams in tremendous action at the height of his career (*Stonex*)

no careless mistake at the wall. He collected himself, measured his stride, drove forward and cleared the wall.

So both were clear yet again. There was a pause. The atmosphere in the stadium, by this time, was as tense as only it can be on these great occasions. Some eight thousand people awaited the appearance of Finality for yet another jump-off.

At this moment the telephone at my commentary position (which is also the control point) rang. My assistant picked up the telephone and then whispered something to me. I was asked to make an announcement. But I did nothing. There was a few moments' pause and then the assistant whispered urgently, 'You were told to give it out'. But I declined. I felt that what I was asked to announce could be made public in a far more dramatic way. Instead, I gave the signal to open the gates from the collecting ring into the arena.

The gates were opened and into the arena rode the mighty Foxhunter—nearly 17 hh—and by his side the little Finality—the David and Goliath that had been thrilling the crowd for the last half hour. When they reached the centre of the arena I gave another signal and the 'boxing lights' were switched on, bathing the whole of the centre of that famous arena in a bright amber light. At that moment Harry Llewellyn leant down and offered his hand to Pat Smythe, who took it. Before that vast crowd they shook hands and Harry Llewellyn took off his hat to Pat.

That gesture made it quite clear that they had decided to divide. Another jump-off would mean that they would either be equal yet again, when already they had jumped quite enough, or one would have to be the loser when neither deserved to lose, so they had decided that they would be equal first. I do not think that I have ever heard a cheer the like of that one which went up in the Harringay Stadium at that second Horse of the Year Show until 1972, (the incident described earlier).

The cheering continued for several minutes as Foxhunter and Finality, and their gallant riders, stayed there in the centre of the arena. Finally, to continuing applause, they rode out, to come back a few minutes later for the presentation of the trophy and the rosettes. Already it was nearly midnight. As far as television was concerned we had over-run by more than an hour, but I doubt if more than a handful of viewers had switched off, for during the last two hours, viewing had been absolutely compulsive—perhaps for the very first time as far as show jumping is concerned.

For more than two hours the telephones were ringing, as viewers called the Harringay Stadium, the Horse of the Year

Show office, the BBC, even my own home, to say what fantastic viewing they had enjoyed. I have little doubt that it was this great competition, which had so caught the imagination not only of the spectators present, but of the millions of viewers watching at home, that was to make the BBC appreciate to the full the possibility of show jumping as a television sport.

More than that, honour was satisfied in every way. The great hero, Colonel Harry Llewellyn, and his immortal Foxhunter had certainly not been disgraced, but Cinderella in the shape of Pat Smythe, with her little mare Finality, had given them a very good run for their money, and had finished on equal terms. This, to the general public, was a fairy-tale ending that could satisfy even the most romantic.

So for nearly thirty years, scarcely a year has passed without some thrilling, momentous event: Harvey Smith tying with himself in the Puissance: the tiny Irish Dundrum thrashing the giants: Britain in mortal combat with Germany: a tenth of a second victory: a tie even. So it goes on year after year, even as recently as 1975, which was not a vintage year for jumps-off. There were two magnificent moments. The first was in the Daily Telegraph Cup, which is an event against the clock in the first round. It was a big course, few were going clear and those who were dared not take risks, so the times were not particularly challenging. Then it was Alwyn Schockemohle's turn. He had won the major event on the first night of the Show, the Butlin Championship, beating David Broome by one second in a jump-off against the clock; at the Royal International in July he had won six major events. On the second night it had been David's turn when he was first, third and fourth in the Phillips Electrical Stakes. Schockemohle had just tipped the last fence in the first round.

Now was his chance to get back into the picture. The time that he had to beat, set up by Ken Pritchard on Longboot, was not all that difficult, but from the way he attacked it, it might have been ten seconds faster. He really raced round that big course, taking incredible risks, never wasting a yard. His time, two seconds faster than Pritchard's, produced a roar from the crowd, for despite his propensity to beat the British, Alwyn Schockemohle is extremely popular in England.

David Broome, by the luck of the draw, was the very next to go. He looked relaxed and a little detached as he always does on the big occasion, giving the impression that he is somehow slightly apart from all the excitement and tension around him. But those who knew him well were familiar with his style, and appreciated as he set off round the arena towards the start that he

The ever-popular Stroller, ridden by Marion Mould (*Roberts*)

really meant business. Deceptively, he appeared as usual to be taking the first fence or two on the slow side; but he was not wasting a second. With Philco perfectly balanced he was able to maintain his rhythm round every corner however sharp: flowing smoothly on often seems more effective than catching up, turning on a sixpence and then accelerating. Half way round the course David's time was exactly the same as Schockemohle's, but on the turn of almost 180° into the last line of fences his smooth cornering gave him the advantage. Flashing down the treble at the end of the course, meeting each element perfectly, he clinched his victory. His time was just .7 seconds faster than Schockemohle's.

The jump-off that had been awaited all the week, however, came on the Friday evening in The Sunday Times Cup. Harvey Smith who had been out of touch throughout the show was the first to go against the clock, on Olympic Star, and he notched up a time of 29 seconds which on nine occasions out of ten would have been unbeatable, even with competitors of the calibre of those ranged against him: David Broome, Graham Fletcher, Malcolm Pyrah: for once the Germans had not survived the first jump-off, both Schockemohle and Fritz Ligges falling by the wayside.

Once again Broome's start was deceptively slow, but once again he kept his mount—this time Sportsman—beautifully balanced and never for one moment lost his rhythm as he swept round from fence to fence. The crowd was now in a fever of excitement. It was obvious that if David avoided having a fence down—and the fences were big ones—he would be very close to Harvey Smith's time. Everyone was trying to keep one eye on the clock and one on David. That this was somehow achieved was evident as the cheer rose in a mighty crescendo when David hurled Sportsman over the last fence to clock the incredible time of 26.8 seconds, over 2 seconds faster than Harvey Smith.

Still, this was not the end. Graham Fletcher with Tauna Dora was obviously determined to lower David's standard. Almost lifting Tauna Dora over every fence he too turned on a sixpence every time, but as he flashed through the finish, the clock showed 27.3: faster than Harvey, but still behind Broome.

Nor was it all over even then, for the fourth finalist, Malcolm Pyrah on the young Severn Valley, really had a crack, despite the fact that his big, impressive bay hardly looked the type of horse to be a speed merchant. Nevertheless, as he turned to the last line it still seemed a possibility that he might beat them all. Unfortunately as he soared at the last he just tipped the top pole and a roar of anguish went up as it fell. His time was within a second of Sportsman's but he had collected four faults.

So David Broome achieved the first of his two hat-tricks of the week, twice winning three major competitions in a row. On the last night with his final win in a major event, this time the Horse and Hound Cup, making his total seven, he managed to beat Schockemohle's remarkable record of six wins during the week of the Royal International Horse Show. Certainly we had seen 'Broomy' at his best. As it had done in the past for so many other great show jumpers, the Horse of the Year Show had inspired him to his greatest.

# 13 'Plus ca Change-'

'Always the same'—absent friends—'Doc' Nichol, 'Handy' Hurrell, David Satow—the course builders—Talbot Ponsonby and Charles Stratton—from Harringay to Wembley—problems of the move—attendance of the Queen—a personal presentation—the royal box—tragedy and the subsequent absence of Colonel Mike—a happy show with a tricky problem

Year after year, scarcely a Show passes without the audience being treated to a really thrilling jump-off. It may be the Puissance with the wall standing at over 7ft: it may be a jump-off against the clock with a tenth of a second victory: it may be a win for one of the crowd's darlings—Harvey Smith, Ted Edgar, David Broome. It may even be a youngster humbling all the big names—John Whitaker, Malcolm Bowey, Nick Skelton, or any of the many oustanding young up-and-coming riders who constantly prove that there is no shortage of riders to step into the existing stars' shoes when they become vacant.

The audience at Wembley can enjoy so much more than just the jumping, having first-hand experience of the extraordinary variety and atmosphere of the show. It is the ten millions sharing the thrills of the Show on the television screen who begin to feel that the show is timeless, as year after year, it provides the expected feast of exciting jumping, brilliant horsemanship, and cliff-hanging competition. There are new names, of course, and each year, a new hero: but there are also the names that have spanned twenty years or more of the Horse of the Year Show: Ted Williams, Alan Oliver, George Hobbs, Ted Edgar, Fred Welch; and, of course, Harvey Smith, David Broome, Paddy MacMahon, Marion Mould and Caroline Bradley. Their names have become household words with millions of viewers, who get to know their horses as though they were stabled up the road.

Though the phrase which is frequently heard, 'the Horse of the Year Show is always the same' can on occasions be derogatory, more often it can be interpreted as a compliment. It is as regular as the Boat Race or the Derby. It can be relied on each October for entertainment of the highest order. Year after year it is peak viewing for millions who can never get to the Empire Pool. For them little change is evident.

In fact, over the years there have been many changes within a framework that since the earliest days has been constant: a case perhaps, of 'plus ca change, plus c'est la même chose' (the more things change the more they are the same). As mentioned earlier, there is now only the smallest handful of survivors from 1949 behind the scenes. But many, either because they have retired or died, who are no longer associated with the Show, have put in many years of service, and contributed greatly to the Show's success. Obviously it is not possible to include every name here, but there are certain people who must be mentioned if the story of the Horse of the Year Show is to be complete.

'Doc' Nichol was chief medical officer from the first Show in 1949 until his death in 1971. Bob Armstrong, familiar to practically every rider who has represented Britain overseas, was the show farrier for many years. Extremely popular and a real driving force was the Wembley general manager John Connell who died still a comparatively young man in 1969, to be succeeded by George Stanton who had first been associated with the BSJA when as chief engineer he was responsible for providing the water jump in the Olympic stadium for the 1948 Olympic Games at Wembley.

Colonel 'Handy' Hurrell has already been mentioned. From the early fifties he had been in charge of all the show classes, becoming an assistant director in 1953. On his retirement in 1974 his tall figure and his friendly, reassuring personality were greatly missed by stewards and exhibitors alike.

A great friend of 'Handy' Hurrell was Major David Satow. His contribution not only to the Horse of the Year Show but to the whole of the horse world has been inestimable. Always cheerful, tireless in his efforts in anything he believed to be worth while, David Satow undertook any job that needed doing, and indeed he filled many positions during his time with the BSJA and the BHS. Although he was at one time clerk of the course at Harringay, at another time a course builder, later press officer, later again public relations officer, finally development officer, it was really as Mike Ansell's right-hand man that he made his greatest contribution, especially in the early years of the Horse of the Year Show. He worshipped Mike Ansell and would do anything for him, though at times he was sorely tried. He would always be the first in the office, in the Bedford Square days, but never grumbled if he was expected to stay late, working on the programme, or correcting proofs, or helping entertain visitors. Mike Ansell had only to mention something he was worried about, and David would dash off to see to it. He was not only Mike's eyes, he was often his front man, protecting him from unnecessary worries which would take up his time, accepting as a

Cavalcade in 1960 with Raimondo d'Inzeo on Merano, David Broome on Sunsalve and Laurie Morgan on Salad Days as the centre piece facing the old Royal Box (*Lane*)

calculated risk that he might be on the receiving end of the boss's wrath should Mike decide that the matter in question was something about which he should have been told.

David Satow has been referred to as 'leg-puller-in-chief'. He loved practical jokes and would go to immense trouble to pull off a joke at someone's expense and on occasions even at the expense of 'Colonel Mike', as all with a military background referred to him. Once, for instance, with George Stanton he arranged for a shower of skins—sheep, goats, anything—to fall on the Colonel's head as he entered the conference room. This, of course, was a way of pulling Mike's leg for employing, as stewards, so many ex-members of his old regiment, the 5th Royal Inniskilling Dragoon Guards, of which he was inordinately proud, and which were popularly known as the 'Skins'.

Latterly, when he had taken up his position as BHS development officer at the National Equestrian Centre at Stoneleigh in Warwickshire, David Satow played a less active part at the Show, but there was always a job for him to carry out with his usual unbounded enthusiasm, right until 1971 when he so tragically died of a heart attack.

There have been more changes in the course building department than any other and as the success of jumping depends very considerably on the building of good courses, they have been of great significance. The first course builder was that great artist Phil Blackmore, who had been brought up in the old BSJA school, indeed, was one of its pioneers. Although he was more at home with the old-fashioned pre-war courses, he was sufficiently broad-minded—which is more than could be said for some of his colleagues—to see the advantages of the new FEI courses. He retired in 1956 having latterly been assisted by Eric Ixer. The following year Colonel Jack Talbot Ponsonby was brought in. An outstanding show jumper himself, the first rider to win the King George V Gold Cup three times, he brought a new dimension to course building, with his notions of related distances, and problem fences, by no means popular with all riders.

Nevertheless, Jack Talbot Ponsonby's courses were respected all over the world; nor can there be any doubt that he was responsible, not only as course builder but as the British team trainer, in improving the standard of show jumping in this country. When in 1969 he died in the hunting field in Warwickshire he had passed on much of his knowledge to John Gross and Alan Ball who succeeded him as course builders. Military duties had prevented Major Guy Wathen who had for a short time been responsible for the courses, from taking on the job. Eric Ixer had emigrated, John Gross was killed driving back from the Royal Cornwall Show—a grievous loss to the BSJA—so it was Alan Ball who became the official course builder in 1970. Clever, conscientious, popular, Alan Ball's courses can hardly be bettered and the fact that riders from all countries like to experience them is evidence of his skill.

One other figure familiar at the Horse of the Year Show for many years was Charles Stratton who acted as Colonel Ansell's escort, piloting him around the arena from dawn to dusk. He had the advantage of being tall, almost as tall as the Colonel himself, which made it easier for him to take Charles Stratton's arm just above the elbow. In recent years Paul Mercer took over this responsible job.

Obviously there must be such changes of personnel, even eventually at the very top. There have been other changes, too, that have been influential on the character of the Show. Undoubtedly the greatest was the change from Harringay to Wembley in 1959.

News that the Harringay arena was to be sold as a storage depot came as something of a shock. We remembered only too well how difficult it had been to find a suitable stadium in 1949. Once again, there was mention of all the old stadiums, Earls Court, Olympia, and so on, but it was the GRA itself who put forward the idea of the Empire Pool, at Wembley. When one realises that even at the time of the last Horse of the Year at Harringay it was not known where the next Show was going to be held one can appreciate how near it came to extinction at the end of its first decade.

It goes without saying that there were plenty of people who said that the Show would never be as good at Wembley as it had

Cavalcade in 1975 with Lucinda Prior Palmer on Be Fair in the centre under the lights (*Lane*)

been at Harringay. Indeed, there are still people, even today, who will tell you that the Show has never had quite the same atmosphere since it left Harringay. But almost certainly this is a case of nostalgia, looking at the past through rose-coloured spectacles as the trials and tribulations, the ultimate triumphs of those pioneering days are affectionately remembered. It is true that there was at Harringay a certain intimacy and certainly in those early days, a very special atmosphere, created by the feeling that all were involved in an exciting new experiment.

However, there can be no doubt that the Empire Pool at Wembley is preferable. Not only is it better situated, with proper car-parking space; not only is it bigger, but it is cleaner, better laid out and from an administrative point of view far better suited to the running of a horse show. There is not a great deal of difference in the size of the arena itself, but the walk-way or surround is far more spacious, allowing for many more stands and creating much less congestion. Wembley, too, has infinitely superior refreshment facilities, in particular at the restaurant overlooking the arena.

There is one significant difference to the arena, and it cannot be denied that it is a disadvantage. At Harringay there were two entrances, one at the end and one at the side. This was obviously a great help in creating an easy flow both in the jumping and showing classes and, particularly in the displays. Although every possibility was explored there is, at Wembley, no way of having a second entrance. With his usual ingenuity, Mike Ansell, having discarded the suggestion that a second entrance or, more accurately, an exit could be created in the corner by the main entrance, turned a necessity into an asset. The entrance was made really imposing with heavy curtains across the front and a spacious collecting area immediately behind. The single entrance-cum-exit makes things far easier for control, and the fact that there is only one entrance was very quickly accepted: though it cannot be denied that as far as displays are concerned, including the parade of personalities and the cavalcade, it would be very much easier if there were a second entrance.

Obviously leaving Harringay after ten years and going to a new home was a big upheaval, especially for the staff and, to a lesser extent, for the stewards. With the usual adaptability and a refusal to consider anything impossible, everyone quickly settled into the new environment. Inspired by Colonel Mike we set about creating an organisation which would be just as efficient as at Harringay.

An outside ring had to be created. Obviously the Olympic stadium at the top of the hill was not suitable as the outside stadium at Harringay had been. A cinder arena was built and received a certain amount of criticism at the time, as it has ever since. It did not have the ideal surface, nor, inevitably, was it very clean. Each year great efforts were made to improve the outside ring, a complete solution only being reached in 1975. Due to the building of the conference centre the stables and outside ring had to be re-sited, so a fine, level sand ring has been created with the stables behind. There is no doubt that the outside ring has been the subject of more controversy, especially with the show class exhibitors, than anything else at the Horse of the Year Show.

At first, on the move to Wembley, the subscribers' lounge and dining room were in the Olympic stadium, but this arrangement found little favour being too far away, and therefore little used. Eventually the solution was to create a marquee with direct access from the Empire Pool. It is certainly better than nothing, but not ideal. It may well be that those who remember it were spoiled by the ideal members' accommodation for the Royal International Horse Show at the White City: two lounges, a dining room, all leading out to the members' blocks of seats and the forecourt bars.

At Wembley all the bars situated round the walk-way have been named after eminent show jumpers. How many are still remembered? Craven A, Earlsrath Rambler, Finality, Jane Summers, Mr Pollard, Nizefela, Silver Mint, Foxhunter, Nugget, Pegasus, Red Admiral, Sheila, Snowstorm, Sunday Morning. The majority, I hope, will still be recalled by many with affection, for they all contributed to the success of the Show in its first decade at Harringay.

The final performance at Harringay on 11 October was for many a sad occasion, but it went through just as any ordinary performance: the musical drive, the dressage display by Willie Schultheis, the Pony Club games, the parade of personalities which that year included the Palomino stallion Bubbly, three of Miss Broderick's lovely Coed Coch ponies, Pretty Polly, the hackney pony stallion Oakwell Sir James and, for the last time, the pit ponies, and the cavalcade. At the end I introduced the Duke of Norfolk, the president of the BSJA who spoke briefly but effectively and was able finally to announce to an apprehensive public that the Show would carry on the following year at Wembley.

It had been hoped that the Queen might attend the final show at Harringay, but it was not possible for her to do so. Although the Queen has seldom missed a Royal International Horse Show, invariably attending on the Wednesday night for the jumping of the King George V Cup, she has only attended the Horse of the Year Show three times since it started, two of those when she was still Princess Elizabeth. This is because the royal family do not, as a rule, return from Balmoral until the week after the Show. Only when affairs of state have brought Her Majesty specially back to London has it been possible for her to fit in a visit to the Horse of the Year Show.

She did come to the 1957 Show, as I have personal cause to remember. It had been decided to commemorate her visit by presenting her with a specially bound copy of a book which I had just had published, entitled *Clear Round, the Story of Show Jumping*, which was the first book of its kind, strangely enough, in view of the mass of equestrian literature that is produced today. Shortly before the Show there had been something of a scare as the parcel containing the book had been mislaid. The Show had already started when it was discovered attached to the notice board outside the secretary's office, presumably someone had found it and hopefully placed it there.

At the beginning of the interval the Duke of Norfolk leaned out of the royal box which was just above control and said he had been told that he had to present a copy of my book to the Queen, but he did not really know what it was all about. I started to explain but he interrupted me by saying that he thought it was best if I did it myself and would I come round? On the way I hurried into the cloakroom to wash my hands, discovering too late that there was no towel. The entrance to the royal box ante-room was just across the alleyway and wiping my hands on my trousers I entered what I expected to be the crowded ante-room. Only the Queen talking to the Duke of Norfolk, two others, and the Duke of Edinburgh standing right against the door, were in there.

Immediately the Duke held out his hand. I had no option but to take it, wet and clammy as my own hand was. He concealed his wince admirably. A few moments later the Duke of Norfolk

A section of the vast crowd reflecting the tension during a major jumping event (*Stonex*)

beckoned me across to the Queen and presented me, but made no mention of the book, which I, in a somewhat embarrassed manner, had to do myself. I gave it to the Queen who thanked me and asked me if I had signed it. When I told her that I had not she asked me to do so, saying: 'Write something nice: it always makes a book so much more interesting.' It is not easy to think of something apt on such an occasion, quite spontaneously. I wrote simply: 'For Her Majesty, on behalf of the British Show Jumping Association and horse lovers everywhere', and was not a little embarrassed to be asked by the Duke of Norfolk to read out to the assembled company, which was now quite large, what I had written.

The re-siting of the royal box was another change at the Horse of the Year Show, though in all probability not one appreciated by the viewing public. It used to be situated half way up one side, on the right as one came through the entrance to the arena. It was very central but it had two disadvantages. The first was that it entailed considerable construction and effectively blocked the narrow alley that ran right down the side of the arena, thus tending to isolate control from the collecting ring area. The second was that to reach the royal box guests had either to go down the main alleyway, which at the time when VIP's were arriving was absolutely jam-packed with people, or to descend to the lower, narrow alley by the ringside, which was not only awkward, but was not exactly an edifying approach to it.

Another disadvantage was that it virtually over-looked control, so that practically anything that was said by control could be heard in the royal box. Of course, this necessitated everyone in control exercising considerable verbal selfdiscipline—which may not have been a bad thing.

I believe that those in the royal box rather enjoyed being able to see all that was going on in control. It is human nature to want to see behind the scenes. Certainly, they enjoyed being able to see the television monitor sets and listen to our commentaries. Prince Philip, on the comparatively few occasions on which he came, always used to sit in the right-hand corner of the royal box so that he could keep an eye on all that was going on in control. In some ways Mike Ansell liked the original position of the box because standing by control he could also have easy contact with those in the box, many of whom were always senior army officers and so previous colleagues of his.

When the Royal International was moved to the Empire Pool in 1970, an attempt was made to make the arena look as different as possible for the Horse of the Year Show. To enlarge the arena, the seats on the ground floor, in front of the narrow alley, were removed. This made it difficult to fit the royal box down the side and so it was moved to the end of the arena, facing the entrance, a far more satisfactory position in every way, particularly from the point of view of access and general viewing. It did mean that the presentation of rosettes and cups had to be done right up at the end of the arena which was not quite so satisfactory, the majority of the audience only seeing the backs of the horses and riders. However, when the winners had lined up in front of the old royal box only half the audience saw their faces. Certainly control heaved a sigh of relief when the royal box was moved and although the move might have been forced upon us of necessity, it turned out to be a great improvement.

On reflection, in nearly thirty years, the Horse of the Year Show has not often been disturbed in the even tenor of its way.

It was disturbed in 1971.

Immediately following the Horse of the Year Show in 1969 the Anglo–Austrian Society put on, at the Empire Pool, a magnificent display by the Spanish Riding School of Vienna under Colonel Hans Handler. A number of the show's senior stewards were asked to help and it is probably true to say that in the end the Horse of the Year Show was responsible for the presentation. It was a wonderful success, with every seat sold out in advance. The final performance was on the Sunday evening. Sir Michael and

Lady Ansell were present, Victoria Ansell looking very frail and tired at the end of a long fortnight. She had been unwell for some time, but despite illhealth insisted always on being at the Show to support her husband and act as hostess.

The following morning she died in her hotel room in London. Mike was devastated. However, in a little over a year, to everyone's delight, he married again. Major General Roger Evans and his wife had been very close friends of the Ansells for a long time. It seemed ideal when one of their sons, Sandy, married the Ansells' daughter, Sarah. Shortly before the wedding Roger Evans died, but on Eileen Evans' insistence the two youngsters were married, as arranged, in November 1968. Not unnaturally, after Victoria's death Mike saw a good deal of Eileen, herself so recently widowed. In December 1970 they themselves were married, and Mike's great happiness was obvious to all. Though Eileen was always at the Shows supporting Mike, one felt that her hopes were really for him to retire and settle down in the country, giving himself more time for the flowers and fishing which he so loved. There is no doubt that his own thoughts were turning in this direction when disaster struck again. At Mere, in Somerset, at the end of August, walking along a pavement with her son Sandy, Eileen was hit by an out-of-control lorry and killed—only eight months after her marriage to Mike.

This time Mike seemed inconsolable and it was not surprising to hear him saying that he would not be able to run the Horse of the Year Show in October, in two months' time. Many people seemed to think that I would have to run it in his absence which worried me. In the third week of September I wrote to him urging him to reconsider his decision, as I felt, first, that to miss the Show would add to his misery; secondly, that to be actively employed in something that he loved doing might help him.

For the normal bereaved person it is meeting people that causes the most embarrassment: looking into their eyes, one sees the grief and pain so clearly that one is embarrassed for them. With a blind person this is not so. A blind person need not meet anyone that he does not actually want to, because there is no opportunity for anyone to catch his eye. At Wembley I felt that Mike could carry on his job without being embarrassed by meeting people who did not know quite what to say to him. If anyone wanted to speak to him, who would cause Mike unnecessary pain, it would be easy enough for his escort to ward him off.

Mike, however, was adamant. He could not face all that would be involved, and so after discussions with senior lieutenants, he asked Captain Jack Webber to act as his deputy. This was a sound decision. Captain Webber was secretary general of the BSJA, whose show to all intents and purposes this was, an immensely respected and remarkably tactful man. He was to be helped by John Stevens who recently had been appointed assistant show director, working permanently at Belgrave Square with Mike Ansell. All stewards, senior and junior, were naturally determined to help in every way they could, determined to make the Show a success, determined not to let down 'Colonel Mike'.

So on 4 October 1971 the Horse of the Year Show opened for the first time in twenty-two years without the name of Colonel Sir Michael Ansell in the programme as show director. But the Show went without a hitch, all involved finding it extremely happy and relaxed, thanks to the particular gifts of leadership of Jack Webber and the foundations laid down over the years by Mike.

There was only one unfortunate episode which caused a considerable ripple on the surface. At the end of one competition Harvey Smith came in for his award on a different horse to the one on which he had competed. This is often done and there is nothing wrong in it. But as he went out, having received the award, he jumped a fence, thus contravening international competition rules (although such an act is permitted in national classes.) The fact that this was the horse on which he was competing in the next event gave rise to suspicion, and, not surprisingly, George Hobbs, chairman of the rules committee of the BSJA objected.

There was a lot of to-ing and fro-ing between the committee, the judges, the rules committee, the problem being that this was a National competition at an international show: and naturally there was a great deal of publicity. Without going into all the rights and wrongs of the case, the perplexities and the personalities, it was not an easy case to handle for the jury, a sub-committee of the Show committee. As Jack Webber was an officer of the BSJA, I had to take the chair. It was unfortunate that the incident should have happened at this particular Show: first because otherwise everything ran so smoothly; secondly, because Mike Ansell was not present.

The committee were not altogether surprised—nor too worried —to be told later by Mike Ansell that he would have handled the whole matter differently and more effectively!

# 14 The Fences Ahead

What of the future?—an ad hoc committee—the inner circle—'leave it to the boss'—what of the chairman?—eventual need for a new constitution—ingredients that have spelt success—the man at the helm—that unique atmosphere—the Cavalcade—'Where in this wide world?'

With the retirement of Colonel Ansell as Show director after the 1975 Show it was obviously the end of an era that had lasted nearly thirty years. Many people understandably wondered whether his departure would noticeably affect the Show for throughout his reign it had been so very much of his own making. Oddly enough, it was only in 1971 that his name appears in the programme as chairman, although he had always taken the chair as show director.

The Horse of the Year Show committee has always been something of an ad hoc, self-propagating affair. It was in 1958 that I was first appointed to the committee, but I have no recollection of being formally elected though I have no doubt that at some committee it was suggested that I should be asked to join it. My name first appears in the programme as assistant director in 1962, but again I doubt if it was ever an official appointment by the committee.

Certainly I cannot recall Mike Ansell being formally elected as chairman of the committee as recently as 1971—indeed I can recall no elections of any sort. Since the beginning 'Colonel Mike' has taken the chair, gathering round him people he thought would be actively helpful or could play a useful role in representing a body of people whose support he needed: for instance, the recent inclusion on the committee of Fred Hartill ensured the support and connivance of owners (Fred Hartill, of course, owns Penwood Forge Mill); the presence of Dick Stillwell, one of the leading trainers of show jumpers, both horses and riders, ensured that top riders felt represented, while Alan Oliver was on hand to look after the rank and file.

Truth to tell, the committee seldom met: frequently not more than once a year, in December or January, to report on the previous Show and consider the estimates and plans for the following year. Normally, the committee would not meet again before the next Show, a 'progress report' from the chairman round

At the end all join in the singing of Auld Lang Syne . . . (*Stonex*)

. . . including the commentators, from the left, Dorian Williams, Christopher Hall, Raymond Brooks-Ward: The Cavalcade can be seen on the monitor set (but not, unfortunately, being transmitted to Britain!) (*Stonex*)

about August sufficing. This may sound somewhat autocratic: but as with many exceptionally able people, committees are not exactly anathema, but they are considered cumbersome and inhibiting. Mike Ansell always preferred to 'use' people, inviting a member of the committee, or possibly two—cronies from the inner circle—to his office at Belgrave Square at about 6 o'clock in the evening. He would pick their brains, discuss matters with them, sound out his own ideas and get their reactions—though it has to be admitted, that with some people, because of the force of his personality, there could be the reactions that Mike wanted, rather than detached, independent ones.

Having talked, considered and absorbed for an hour or two over a couple of scotches Mike would then thank whoever it was he had invited, and persuade them, without difficulty, that it was going to be 'the hell of a Show', infecting them effortlessly with his own enthusiasm. Next day, or a short while later, he would 'phone other members of the committee and inform them that so-and-so and so-and-so had agreed with him that such-and-such a thing would be a good idea and that he was sure that they would agree too. They usually did.

With a chairman of his personality this method worked extremely well, especially as he was also Show director. As is so

often the case with great men, even more so with blind people, Mike has always had an instinctive ability to know what people are thinking or saying, without actually being told. He would be aware that amongst certain people there was a notion that a certain change should be made; or that someone in a certain position was rocking the boat; or that there was in some particular department inefficiency. He would therefore find an opportunity to bring up the matter, agreeing or disagreeing, but at least giving it an airing, thus making sure that people could not complain of their ideas or complaints being stifled or ignored.

Certainly the majority of the committee—and in any committee, even in a non-meeting one, there are always a certain number of passengers—felt that they made what contribution they could and were quite content to leave major decisions to the chairman for whom it was virtually a full-time job.

Understandably, there was a reluctance for a member of the committee to be too dogmatic on any point. For them the Horse of the Year Show was very much a part-time affair, each member having a full-time job of his own. Mike Ansell could always remind them as he not infrequently did, that whereas they had only been able to give sporadic thought to the problem he had been able to give it the whole of his attention. Often he has pointed out that, coming up in the train from his home in Devon, while other people are reading their papers, or looking out of the windows, or opening up a conversation with the person opposite whose eye they have caught, he, being blind, has just sat silent, thinking. For all that part of the day when he is not actually engaged in conversation he is *thinking*: probably spending ten times as much time on a problem as any other member of the committee.

His contribution, therefore, has been enormous. Can it ever be equalled? Can anyone else give even a fraction of the time to the Show that he has? It is worth quoting from a perceptive article in *The Times* on the announcement of his retirement:

> He has been chairman of the BSJA for two decades, of the British Horse Society for nearly as long, director of the Royal International Horse Show and Horse of the Year Show, and is now chairman of the body which supersedes all else, the British Equestrian Federation.
>
> To the world of show jumping he is a hero. Surrounding the seat of power, the young and middle-aged men who like running international shows have been welded into a team of junior officers characterised by the sort of unquestioning loyalty which obtains in house sides or regiments. Arbitrary decisions are made and obeyed, and the unit goes from strength to strength under the leader.

It may be that with his retirement from the two shows' directorships and the end of his term of office as chairman/president of the British Equestrian Federation at the beginning of 1976, he will still like to be involved in the horse world in some way: just how remains to be seen. What is certain is that he will no longer be director of the Horse of the Year Show.

There can be little doubt, however, that for the foreseeable future the Show will carry on very much as hitherto. In the first place his senior stewards are a trained team who for many years have worked under him and with him. Many of them, as described in an earlier chapter, have made very considerable contributions themselves: indeed, have played a part without which the show could not possibly have been the success that it is.

In John Stevens there is a show director who has proved himself an able, efficient and extremely conscientious lieutenant, who has worked closely with Colonel Ansell for a number of years and has been responsible for all the details, in every department. He was, of course, acting show director, under acting chairman Jack Webber, in 1971, when Colonel Ansell was absent and when the Show ran extremely smoothly and happily.

Admittedly many of the senior stewards are not as young as they were. Mike Ansell was in his early forties when he took over the running of the Horse of the Year Show. Few, if indeed any of those left in positions of responsibility are as young as that.

The immediate question is whether the high standard of the Show can be maintained once it is deprived of the services of its original presiding genius. As show director and chairman Mike Ansell has been a unique personality who has given unceasingly from his great reservoir of genius, has supplied unlimited drive, provided unchallenged inspiration. It would indeed be a miracle to find a like character with the same time and energy to devote almost exclusively to the shows: not impossible, but unlikely. It is obviously going to be essential to share the responsibilities and spread the driving force required to run the Show. Eventually, perhaps, as the old associates of the founder disappear from the scene, there will appear a new personality with a similar character and flair, though in my opinion, the Show's continued existence does not depend upon the emergence of such a character. After all the Derby changes little from year to year whoever may be responsible for all the organisation behind it. The same could be said, more appropriately perhaps, of such an occasion as the Lawn Tennis Championships at Wimbledon, or the Regatta at Henley. Yet it is likely that those most closely involved in these events, would claim that much depends upon the hand on the helm: and Mike's hand has been on the helm of the Horse of the Year Show for a long time.

It is indisputably the end of an era. The new era, inevitably, will bring about changes. But, the unique character of the Show will surely survive. Of this there can be no doubt.

Just what is it that has made the Horse of the Year Show so different, has given it this unique atmosphere, created for it such a remarkable reputation?

To summarise what has, in fact, been the gist of this book: first there is the variety. No show before, or indeed today, can boast the extraordinary variety of the Horse of the Year Show. In one performance there are at least six different ingredients, jumping, driving, gymkhana events, dressage, parades, show classes. Secondly there is the time-keeping. In Britain today one is inclined to take it for granted that horse shows run to time—though, in fact, by no means all of them do. That most do is almost entirely due to the example of the two Shows run since the war by Colonel Mike Ansell: good time-keeping is now considered essential. Regrettably, many continental shows have yet to discover this necessity. In this country, more than on the continent, punctuality is disciplined by television, the need to finish a competition before the end of transmission.

Thirdly, of course, there is the unusually slick presentation—something more associated with the theatre than with a horse show and still more associated with the army. Colonel Ansell is, as is obvious to anyone associated with him, a dedicated soldier. For him organising a show is like planning a campaign, with the same battalion orders, bumph, delegation of responsibilities, discipline. Before the war, shows were either run by the military, closely associated with the army, probably at a garrison town, or they were run by a country community, as in the case of a small local show. The former was efficient, well organised, exclusive. The latter was something of a garden party: informal, leisurely, seldom of a very high standard. There were a few large agricultural shows and county shows, run, of course, by amateurs—traditional, enjoyable, popular, largely because of their rarity. There were even fewer society shows—the International at Olympia, the Richmond Royal Horse Show—professionally run with the emphasis very much on the social side and on glamour.

Naturally the war swept away most of these in their original form. When shows started again after the war, not surprisingly, there was considerable dependence on retired army officers. Nowhere was this more evident than at the White City when the International was revived in 1947 and at Harringay when the Horse of the Year Show was founded in 1949. Many army colleagues of Mike Ansell were involved in both Shows.

It has become a habit in recent times to mock or criticise 'the Colonels', yet it cannot be denied that in the horse world, both as organisers and, in another sphere, as instructors, they have made a great contribution. Certainly the efficient running of many shows, not least the Horse of the Year Show, owes much to the military background of those responsible for the organisation. Undeniably it has been very much a feature of the Horse of the Year Show, responsible as much as anything for earning it the reputation of the most slickly produced Show in the world.

That one man, has been responsible for the overall success of the Show this book has clearly shown: for him the Show has been a challenge, its success a goal that had to be achieved, even at times ruthlessly, but it has also been fun. 'It's all such fun', he has said a thousand times: but not only fun for himself. It has also been fun for the many who have helped him, and for all those who, taking part as exhibitors or attending as spectators, have benefitted from his leadership.

Much of the Horse of the Year's appeal is in the wonderful audience involvement that the Show has always inspired. This is due to two of its unique features. One is the proximity of the audience to the arena. In an outdoor show there is, inevitably, a certain remoteness. In many of the big indoor shows on the continent the halls where the shows are held are so vast that there is a certain unreality. But even from the highest seat in the stadium one feels that one is right on top of the horses. The lighting heightens this feeling. Those lower down almost feel that they are jumping with the horses or that the horses are just about to jump into their laps. For many people horses in action have never been so close.

The other is this end-of-term atmosphere referred to earlier in the book. Although indoor shows are now so much a feature of the winter months, so that most of the riders will continue to be competing against each other—even though they may be riding different horses, probably young ones that they are taking the opportunity of bringing on—yet the Horse of the Year Show is still considered to be the climax of the season. Many in the audience, even those not particularly closely associated with the sport seem to become aware of this, which gives the Show the feeling of a special occasion. It is as though everyone is in festive spirit.

Even those attending the Show for the first time—as is the case with thousands—are, it seems, quickly infected. For many others, the majority, perhaps, the Show is an annual occasion that they would be reluctant to miss. More than at any other show of which I have experience there is a very real bond between performers and spectators. It is something that was common in the old music hall days, but must be almost unique in a horse show.

Final fanfare (*Stonex*)

This attitude is both warming and inspiring, and, even after all these years of familiarity it is still—for me at any rate—wonderfully exciting. The whole stadium is charged with this almost indefinable atmosphere that has made this show different to any other horse show in the world.

Nor would it be wholly true to suggest that this atmosphere developed slowly over the quarter of a century and more of the Show's history. As I have made clear in an earlier chapter the atmosphere was present at that very first show. Within two or three years it had created a reputation which was the envy of many a very much longer established show. The last night was already regarded as the highlight of the show jumping year, being fully booked within hours of the box office opening.

It is, of course, on that last night that one gets the feeling of all those involved being brought together as one great corporate whole. Every facet of the horse world is represented. They are united as one in the magnificent cavalcade assembled to pay its final tribute to the horse.

The trumpeters of the Household Cavalry enter in darkness, are suddenly hit dramatically by the spotlights in the centre of the arena, blow a fanfare and take up their positions forming a gateway at the entrance to the arena. Each group is then announced and enters to its own music. First the heavy horses to one of the tunes in their display; next the Pony Club teams to *Boys and Girls Come out to Play*; next the personalities to their exit tune *My Hero*; the Riding Club quadrille of the year (the four finalists compete at different performances of the Show) to their own tune; then the horses of the year; the champion hunter, champion hack, champion cob and so on, followed by the winners of the national jumping events, the national ladies' champion, the national young rider champion and so on, with the leading show jumper of the year, some twenty in all. For them the band plays *There's no Business like Show Business*. Next come the riders from overseas, perhaps four different teams; for them it is *Will Ye no' come back Again?*

Finally there are the British international riders: they might all be riding as individuals or as members of winning teams, the three-day event team, the junior show jumping team, winner of the President's Cup (the international team championship). There may well be as many as twenty different riders, all of whom have represented Britain. Appropriately their music is *There'll Always Be an England*. From all of these a centrepiece is selected, the star of that year's cavalcade; it may be an individual, or a winning team, or perhaps some special personality such as Arkle or Mandarin. As arranged at the rehearsal on foot at 5.30 that afternoon—the only rehearsal there is—everyone has filed to his or her alotted place, with the star at the centre.

'The cavalcade is now assembled'.

The lights are dimmed, spots only brilliantly illuminating the Trumpeters of the Household Cavalry and the centrepiece. Despite the packed stadium and the 120 or more horses and ponies in the ring, there is a remarkable, expectant hush, broken only by a horse impatiently pawing the tan or the clink of a spur or the jingle of a bridle.

This, by tradition, is the moment when we pay our special tribute to the horse for all the pleasure that he can still give us, even in an age of machinery and supersonic flight. It was in 1954 that Mike Ansell, travelling down to his home in Devon, met Ronald Duncan who was going down to his farm in the West Country. Knowing that Duncan was a distinguished poet and playwright—his play in verse, *Don Juan*, is a classic—and knowing that I had been hoping for some time to find just the right 'quote' with which to finish the Show, rather than have to rely on my own inspiration, he suggested that he write something brief and appropriate, an apt tribute.

Before the end of the journey Ronald Duncan had composed his famous lines. Though not in verse they have the elegance of a sonnet, and even after all these years they never fail to move. To omit them now, at the end of the Show, would be as unthinkable as dropping *Rule Britannia* from the last night of the Proms.

The Tribute has often appeared in print, but, as with the Show itself, it must bring to a close this book on the Horse of the Year Show.

> Where in this wide world can man find
> nobility without pride, friendship without envy,
> or beauty without vanity? Here, where grace is
> laced with muscle and strength by gentleness confined.
> He serves without servility, he has fought
> without enmity. There is nothing so powerful,
> nothing less violent; there is nothing so quick,
> nothing more patient.
> England's past has been borne on his back.
> All our history is his industry. We are his heirs,
> he our inheritance.
> *The Horse!*

# Index

Numbers in italic refer to illustrations

*Agadir*, 135
Allen, Major Bill, 78
Allen, Brigadier John, 22, 26
Anne, HRH Princess, *93*
Ansell, Colonel George, 18, 89
Ansell, Colonel Sir Michael, 11, 15, 16, 17, 18, 20–26, 30, 35, 41, 47, 58, 60, 63, 66, 71, 76, 77, 82 et seq, 89 et seq, 100, 103, 107, 110, 111, 114, 145, 146, 147, 149, 152, 153, 154, 155 et seq, 165, *90, 93, 97*
Ansell, Lt Colonel Nicholas, 91
Ansell, Victoria, Lady, 91, 92, 153
*Arkle*, 53, 164, *47*
Argyll and Sutherland Highlanders, 45
Armstrong, Bob, 145

Ball, Alan, 11, 73, 85, 88, 147, *83*
Barker, David, 25, 103, 111
Barnes, Gerald, 63, 68, 103
Barnes, Janet, 106
Barnes, Mary, 63, 103
Barnes, Sheila, 63, 103
Barnes, Sylvia, 63, 103
Barnes, Tom, 63, 103
Beaufort, Duke of, 38, 39, 41
*Be Fair*, 148
Black, Bob, 111
Blackmore, John, 58
Blackmore, Phil, 24, 146
*Black Magic of Nork*, 48
*Black Prince*, 57
*Blarney Stone*, 29
*Blinkers*, 47
*Boomerang*, 34
*Bossy*, 42
Bourne, David, 65, 68, 84
Bowey, Malcolm, 144
Bradley, Caroline, 144, *59, 86, 119*
BBC TV, 17, 66, 76, 87, 88, 120 et seq
British Equestrian Federation, 87, 93, 158, 159
British Horse Society, 9, 18, 19, 20, 21, 24, 35, 71, 93, 145, 146, 158
British Show Jumping Association, 20, 21, 22, 23, 24, 25, 26, 33, 35, 37, 42, 65, 93, 145, 146, 153, 154, 158
Broome, David, 15, 17, 28, 125, 133, 134, 135, 141, 142, 143, 144, 146, *33, 59, 124*
Brooks-Ward, Raymond, 65, 88, 103, 117, *156*
Brunt, Leonard, 59, 60, 68
*Bubbly*, 50, 150
Bullen, Colonel Jack, 100

Bullen, Jane, 103
Bullen, Jenny, *see* Loriston-Clarke
Bullows, *see* Wright
Bunn, Douglas, 25, 103
*Burrough Hills*, 48
Butler, Brian, 25

Cancre, Michèle, 20, 24, 26, 28, 29, 85
Carter, Len, 25
Carver, Mrs L., 25, 48
Carruthers, Mrs Pam, 25
Cat and Custard Pot, The, 47, 99 et seq
Cavalcade, 17, 29, 53, 120, 123, 135, 146, 148, 150, 164, 165, 156
Charmartin H., 110
Clarke, Bill, 25
Clear Round, 150
Collings, Captain Tony, 18, 21, 22, 35, 37, 41, *22*
Connell, John, 59, 145
Corry, Colonel Dan, 91
*Costa*, 132
*Count Dorsaz*, 48
*Count Orlando*, 48
*Craven A*, 28, 50, 132, 149
Creswell, John, 25
Creswell, Michael, 106
*Cruachan*, 45
*Crudwell*, 56
Cubitt, Colonel the Hon Guy, 22, 94
*Curioso*, 111

*Daily Mail*, 117
Daily Telegraph Cup, 141
*Daisy*, 47
Darragh, Paul, 26
Dean, Captain J. F., 48, 77, 78
Dean, Mrs Lilian, 72
Dean, R. W. (Bob), 71, 72
*Deauvillois*, 111
*Der Lord*, 135
*Devon Loch*, 25, 48
d'Inzeo, Colonel P., 69, *78*
d'Inzeo, Major R., 69, 146
de Selliers de Moranville, 24
d'Oriola, J., 20
d'Orgeix, Chevalier, 20, 24, 26, 28
Dimmock, Peter, 127
*Doneraile*, 133
*Dracup, PC*, 97
Duncalf, Bill, 128
Duncan, Ronald, 29, 97, 165
*Dundrum*, 141
Dungworth, Len, 22, 24

*Earlsrath Rambler*, 25, 149
*Easter Parade*, 26
East Grinstead Riding Club, *113*

Edgar, Ted, 25, 63, 80, 144
Edinburgh, HRH Prince Philip, Duke of, 95, 111, 114, 150, 152
Eldonian, 109, 112
*Emily Little*, 45
Enfield Chase Pony Club, 117, *113*
*ESB*, 25, 48
Evans, General Roger, 153
Evans, Mrs Eileen, 153

FEI, 26, 107, 146, 147
*Field, The*, 121
*Finality*, 28, 29, 121, 137, 138, 140, 141, 149
Fischer, G., 110
Fleming, Mrs S., 25
Fletcher, Graham, 17, 142, 143
*Flipper*, 79
Fossey, Yvonne, 25
Fox, Paul, 131
*Foxhunter*, 17, 19, 25, 26, 29, 48, 73, 88, 121, 138, 140, 141, 149, *19*
Francis, Dick, 56
*Freebooter*, 53

*Galway Bay*, 25
*Gay Donald*, 47
*Gay Lady*, 28
Gentle, Frank, 21, 24, 35
George, B. E., *47*
Gilhuys, Henri, 110, *74*
Glendenning, Raymond, 106
Goddard, Harry, 103
*Go Lightly*, 24, 132
Grayston, Bob, 24
Grayston, Jack, 24
*Great Moments in Sport: Show Jumping*, 138
GRA, 21, 23, 29, 35
Griffith, David, 59
Gross, John, 147

*Hack On*, 24
Hague, Paddy, 103
Halifax, Earl of, 38, 39
Hall, Christopher, 65, 68, 88, *156*
Hammond, Ann, 25
Handler, Colonel Hans, 152
Hanson, Bob, 22, 38
Hanson, Bill, 11, 38, 87
Harringay, 19, 21, 26, 28, 29, 30, 32, 36, 38, 48, 100, 107, 114, 121, 135, 137, 140, 145, 147, 148, 149, *44*
Hartel, Lise, 107
Hartill, Fred, 155
Haydon, Cynthia, 103, 111, *54*
Hayes, Seamus, 132, 133, 136, 137
Hayter, Rosemary, 103

Heavy Horses, 15, 31, 57, *16, 31, 36, 104*
Hickstead, 25, 135
*High and Mighty*, 48
High Peak Pony Club, *118*
*Highstone Nicholas*, 50, *49*
Hindley, Reg, 22, 25, 29, 38
Hobbs, George, 25, 103, 134, 144, 154, *136*
Holland-Martin, Ruby, 22, 25
*Hollywell Florette*, 42
Horse and Hound Cup, 28, 143
Household Cavalry, 45, 48, 80, 164, 165, *163*
Howes, Ray, 134, 135
Hudson, Tom, 65, 68
Hurrell, Colonel 'Handy', 64, 68, 145
*Hyperion*, 53

Inderwick, Mrs Sheila, 25
Ivens, S., *39*
Ixer, Eric, 147

Jane Summers, 149
Johnsey, Debbie, 28
Jollye, L., 59
Jorrocks display, 54, 134
*Jubilee*, 107
Junior Leaders Regiment, 10, 73, 78, 88, 130, *9*

*Kadett*, 48
*Kalkallo Prince*, 135
Kellet, Iris, 26
Kent, Derek, 103
Kidd, Hon Mrs Janet, 106
Kidd, Jane, 106
Kidd, John, 106
*Kilbarry*, 47
*Kilmore*, 53
King George V Gold Cup, 19, 25, 73, 134, 135, 138, 147, 150
King's Troop, Royal Horse Artillery, 45
Koechlin, *see* Smythe
*Knobby*, 38
Krier, M., 26, 107

Lane, Bay, 25, 28
Leading Show Jumper of the Year, 28
Leckhampstead Surprise, 42
Lee, Sgt Major, 100, 103
Lee-Smith Wightman, Mrs, 25
*Legend*, 38
Le Jumping, 19, 20, 21, 27
*Leopard*, 89
Letherby and Christopher, 67
*Liberty Light*, 38
Ligges, Fritz, 142
Lillingstone Again, 42
Livingstone-Learmouth, Len, 67
Llewellyn, Colonel Harry, 19, 22, 25, 26, 28, 29, 48, 73, 121, 138, 140, 141, *19*
Llewellyn, Mrs Harry, 26
Lloyds Bank In-Hand Championship, 39

Lorke, Otto, 107
*Longboot*, 141
Loriston-Clarke, Mrs Jenny, 48, 103

Macken, Eddie, 17, 26, *34*
MacMahon, Paddy, 28, 102, 135, 144
McHugh, Mrs Jane, 111
Macintosh, Mrs Irene, 25
McKenna, Mrs, 67
Makin, T., 28, 29, 132
*Mandarin*, 53, 164
Marmont, Ronnie, 103
*Marius*, *119*
Marsh, Sam, 103
Mason, Diana, 25
Massarella, A., 25
Massarella, J., 25
Meade, Richard, 28
*Merano*, 146
*Merely a Monarch*, 50
Mercer, Paul, 82, 147
*Meteor*, 107
*Mighty Fine*, 29, 38
*Mr Pollard*, 149
*Mister Softee*, 50
*Monty*, 132
Moore, Ann, 25
Moore, Joe, *109*
Morley, Ann, *136*
Morgan, Laurie, 146
Moss, Pat, 25
Mould, Marion, 63, 142, 144
Mouncer, Alan, 130
Muir, Anne, *51*

National Equestrian Centre, 64, 146
*New Yorker*, 86
*Niblick*, 28, 132
*Nickel Coin*, 42, 50
Nichol, 'Doc', 25, 145
Nichol, Betty, 25
*Nicholas Silver*, 53
Nicholson, David, 111, *105*
*Nizefela*, 25, 53, 137, 149, *46*
Norfolk, Duke of, 150, 151
*Norwood Unique*, 38
*Nugget*, 136, 137, 149

O'Brien, Joe, *57*
Oliver, Alan, 22, 25, 28, 106, 132, 144, 155, *19, 27*
Oliver, Paul, 106
Oliver, Vivien, 106
Oliveira, Nuno, 110, *70*
Olympia, 20, 80, 103
*Olympic Star*, 142
Orlandi, Vittorio, 60
Orssich, Count Robert, 25, 106
Orssich, Susan, 106

Palethorpe, Jill, 25
Parry, Elsa, 72
Payne, A. H., 22, 132
Pearson, W. S., 25
*Pegasus*, 50, 149
*Pele*, 26

*Penwood Forge Mill*, 135, 155, *102*
Peralta, Angel, 107, 110
Personalities' Parade, 16, 42, 46, 53, 56, 57, 85, 149, 164, *43*
*Philco*, 142
Pierce, Stella, *see* Carver
Pinches, John, 71
*Pioneer*, 78
Pit Ponies, 45, *44*
*Planet*, 132, 133
Ponies of Britain, 107
Pony Club, 13, 15, 26, 47, 48, 50, 85, 94, 99, 129, 149, 164, *13, 113, 115, 118*
Porlock Vale Riding School, 41
*Pretty Polly*, 45, 48, 150
Pride, Barry, 68
Pride, Roger, 68, 84
Prior-Palmer, Lucinda, 148
*Prince Hal*, 28, 132
Pritchard, Ken, 141
Pyrah, Malcolm, 17, 142, 143

Queen Elizabeth II, HM the, 42, 48, 150, *151*
Queen Elizabeth, the Queen Mother, 48

*Red Admiral*, 22, 149
*Red Knight*, 132
*Red Star*, 132
Reynolds, General Jack, 85, 87
*Rex the Robber*, *34*
Robeson, Peter, 25, 28
Rochford, Lulu, 25
*Rosina Copper*, 45
Royal Armoured Corps, 10
Royal Army Service Corps, 50, 77
Royal Corps of Transport, 10, 48, 78
Royal Inniskilling Dragoon Guards, 5th, 65, 88, 89, 91, 100, 146
Royal International Horse Show, 18, 20, 21, 23, 29, 36, 42, 58, 68, 72, 78, 80, 92, 93, 137, 141, 143, 149, 152
*Royal Lord*, 134, *136*
*Russian Hero*, 50

*Salad Days*, 48, 146
*Salvador*, *120*
*Sammy Dasher*, *39*
Sargent, Sir Malcolm, 31, *133*
Satow, Major David, 25, 145, 146, *24*
St Cyr, H., 107
Schockemohle, Alwyn, 17, 34, 69, 135, 141, 142
Schockemohle, Paul, 69, 135
Schulteis, Willie, 107, 110, 149
*Sea Prince*, 24
*Severn Valley*, 143
*Sheila*, 132, 149
*Silver Mint*, 42, 66, 87, 149
Simon, Hugo, 79
*Simona*, 134
Skelton, Nick, 144

Smith, Harvey, 17, 80, 137, 141, 142, 143, 144, 154, *59*, *120*, *125*, *126*
Smythe, Pat, 25, 28, 137, 138, 140, 141, *27*
Snodgrass, J., 28, 138
*Snowstorm*, 29, 149
*Specify*, 50
Spanish Riding School, 111, 152, *52*, *116*
Spooner, Mrs Glenda, 107, 110
*Sportsman*, 143, *33*
*Spring Meeting*, 28
Springer, Frau, 110
Stanton, George, 59, 88, 145, 146
*State Visit*, 37
Stead, John, 103
Steele-Bodger, H., 25
Steenken, Hartwig, 134, 135
Stevens, John, 58, 83, 84, 85, 87, 154, 159
Stewart, Colonel Duggie, 121
Stilwell, Dick, 111, 155
Stovold, Ray, 72
Stratton, Charles, 82, 147
Street, General Vivian, 82, 94
*Stroller*, 50, *142*
Styles, Maureen, 25
*Sucre du Pomme*, 26
Sumner, Hugh, 29

*Sunday Morning*, 149
Sunday Times Cup, 87, 88, 142
*Sunsalve*, 133, 134, 146

Taafe, Pat, 53, *47*
Talbot-Ponsonby, Colonel Jack, 147
Tatlow, David, *37*
*Tauna Dora*, 143
Taylor, Tom, 25
*Teal*, 45
*Team Spirit*, 53
Theidemann, Fritz, 107
*Tim*, 28
*Times, The*, 158
*Tit Willow*, 48
*Tosca*, 28, 50, 53
*Tribute to the Horse*, 97, 165

Vale of Aylesbury Steeplechase Display, 103, 106
*Vibart*, 50
Vine, David, 129
Viner, Fred, 130
Von Nagel, Baroness Ida, 107

Walker, Major Desmond, 77
Walmsley, John, 136, 137
Walwyn, Fulke, 53

Wathen, Colonel Guy, 82, 147
Waterloos, Willie, 24
*Waving Corn*, 136, 137
Webber, Captain Jack, 24, 65, 153, 154, 159
Welsh, Fred, 103, 144
Westminster, Anne, Duchess of, 53
Whewell, Gene, 28
*Whisky and Splash*, *10*
Whitaker, John, 144
White City, 18, 19, 21, 23, 25, 26, 29, 36, 92, 93, 149, 160
White, Wilf, 25, 53, 121, *19*, *46*
Whitehead, Mary, 25
Whitehead, Mrs 'Pug', 99
Willcox, Sheila, 48
Williams, Mrs Brenda, 25
Williams, Ted, 28, 144, *12*, *59*, *139*
Williams, Colonel V. D. S., 18, 21, 22, 89, 93
*Winston*, 42
Winter, Fred, 53
Wofford, Warren, *59*
Woodhall, Syd, 25
Woollam, J., 28
Worboys, Major George, 58
Wright, Lady, 25
*Wyndburgh*, 53